AVID

READER

PRESS

ALSO BY ERIK HOEL

The Revelations

The World Behind the World

CONSCIOUSNESS, FREE WILL, AND THE LIMITS OF SCIENCE

Erik Hoel

Avid Reader Press

NEW YORK   LONDON   TORONTO   SYDNEY   NEW DELHI

AVID READER PRESS
An Imprint of Simon & Schuster, LLC
1230 Avenue of the Americas
New York, NY 10020

Illustrations are courtesy of Julia Buntaine Hoel—except images on page 87 are from Louisa S. Cook, *Geometrical Psychology, Or, The Science of Representation: An Abstract of the Theories and Diagrams of BW Betts* (London, George Redway, 1887).

First Avid Reader Press trade paperback edition August 2024

AVID READER PRESS and colophon are trademarks of Simon & Schuster, LLC

Simón & Schuster: Celebrating 100 Years of Publishing in 2024

For information about special discounts for bulk purchases, please contact Simon & Schuster Special Sales at 1-866-506-1949 or business@simonandschuster.com.

The Simon & Schuster Speakers Bureau can bring authors to your live event. For more information or to book an event contact the Simon & Schuster Speakers Bureau at 1-866-248-3049 or visit our website at www.simonspeakers.com.

Manufactured in the United States of America

1   3   5   7   9   10   8   6   4   2

Library of Congress Cataloging-in-Publication Data

Names: Hoel, Erik, author.
Title: The world behind the world : consciousness, free will, and the limits of science / Erik Hoel.
Description: New York : Avid Reader Press, 2023. | Includes bibliographical references. | Summary: "From Dr. Erik Hoel, The World Behind the World delves into the quest for a theory of consciousness that will trigger a paradigm shift in neuroscience and beyond"—Provided by publisher.
Identifiers: LCCN 2023011671 (print) | LCCN 2023011672 (ebook) | ISBN 9781982159382 (hardcover) | ISBN 9781982159399 (paperback) | ISBN 9781982159405 (ebook)
Subjects: LCSH: Consciousness—Research—History. | Neurosciences—History.
Classification: LCC BF311 .H6145 2023 (print) | LCC BF311 (ebook) | DDC 153—dc23/eng/20230415
LC record available at https://lccn.loc.gov/2023011671
LC ebook record available at https://lccn.loc.gov/2023011672

ISBN 978-1-9821-5938-2
ISBN 978-1-9821-5939-9 (pbk)
ISBN 978-1-9821-5940-5 (ebook)

To Roman Hoel

May he and his generation go further

than I and mine could

Contents

The World
Behind the World

CHAPTER I

Humanity's Two Perspectives on the World

We are all of us dualists. It cannot be helped. Both in culture and in our own personal lives, human minds are treated as qualitatively different in kind than machines. Cars break down due to clear mechanical reasons, but when human minds break down, it involves complicated psychological explanations—invisible tidal forces we cannot point to or easily visualize. We know that all of us live within our own solipsistic stream of consciousness filled with internal thoughts and feelings and experiences. It is only in moments of ecstasy or intense physical activity that our minds and bodies seem inseparable—otherwise, the two can be as distant as can be. Each of us knows what it's like to wash the dishes and recall an embarrassing moment in high school that no one else ever bothers to remember. Because of this constant churning internal stream we can in one instance be magnanimous to a stranger and, in another, a selfish devil, all without warning. Emotions whip and tear at us, and sometimes we yell at those we love, or panic in the worst moment, or are overwhelmed into an unwise proposal of great consequence.

While the richness of our consciousness, what it is like to be us, overflows our ability to express it, we have a language around consciousness that allows us to fluently talk about minds. We regularly refer to thoughts, feelings, memories, inclinations, emotions, sensations, perceptions, confusions, illusions—these are not just the building blocks of our daily lives and the minutiae of our streams of consciousness, but also the material out of which the greatest artists and writers make their art. A modern human is fluent in these concepts, able to deploy them to discuss their friends, their family, their enemies, themselves.

This language of mind is based on taking the *intrinsic perspective*. It's the frame we take on when discussing the events that occur only within the mansions of our minds. It's a way to portray, represent, make sense of, and even manipulate mental activity. It is in literature that the intrinsic perspective reaches its apotheosis. When James Joyce introduces the protagonist of *Ulysses*, Leopold Bloom, he writes:

> Leopold Bloom ate with relish the inner organs of beasts and fowls. He liked thick giblet soup, nutty gizzards, a stuffed roast heart, liverslices fried with crustcrumbs, fried hencods' roes. Most of all he liked grilled mutton kidneys which gave to his palate a fine tang of faintly scented urine.

We can easily picture the extrinsic actions here, Bloom guzzling down a veritable menagerie over his plate. Already we have a sense of the man, who has a bit of the glutton in him. But how do we picture the intrinsic aspect of Joyce's description of Bloom? The "fine tang" of the mutton kidneys? Where can we point to the "fine tang" occurring? His tongue? There is nothing we can picture with regards to it, no action that accompanies it, no verbal report by Bloom. And yet we can imagine what the mutton *tastes* like, we can even imagine what it is like to be Bloom—not only that, we can conceptualize how one might like a taste that is vaguely reminiscent of something disgusting, that it would give the taster the smallest thrill. We know that when it comes to our pleasure, sometimes it is more artful for a sensation to have a bit of provocation or discord, rather than merely syrupy sweetness. The intrinsic perspective involves an understanding of the subtleties of mind like this, and the ability to imagine what it is like to experience something is an ability that we spend a lifetime developing.

Sometimes we take the intrinsic perspective on things we shouldn't. Anyone who has talked to a car making a funny sound is guilty of this. "Why?" we ask, fruitlessly, cajoling and bargaining, all to no avail. The mechanic, in turn, simply finds the piece that is broken. For it is obvious that the intrinsic perspective is not useful to apply beyond humans and our close animal compatriots.

Instead, when we look out at nature, humanity has found it immensely useful and informative to take the polar opposite of the intrinsic perspective: the *extrinsic perspective*. Taking the extrinsic perspective on the world means viewing it as consisting of machinery, mechanisms, formal relationships, extension, bodies and elements and interactions. Wheels within wheels, all the way down to atoms. The extrinsic perspective comes into play whenever we use a tool, or navigate the world, or try to understand how some system causally functions, from the most complicated economics to the simplest series of pipes. It is in science that the extrinsic perspective reaches its apotheosis. Lucretius, in his poem "On the Nature of Things," published in the first century BCE, gives perhaps the most prescient historical account of the extrinsic perspective that would eventually become so central to science:

All nature, then, as self-sustained, consists
Of twain of things: of bodies and of void
In which they're set, and where they're moved around.

Of course, the extrinsic perspective has evolved since Lucretius's time—now we regularly speak of waveform collapses, of kinetic motion, of stellar nurseries, of mucous membranes, of hydraulic pressure, of melting temperatures, of tensile strength. The extrinsic perspective has matured into a penetrating and telescopic view of the world. And not just the outside world. For we also speak of our own cells and genes and proteins within. We apply the medicine we are given by the mechanics of our bodies. We swallow the pills. And often the pills work, and we are left with the impression that there exists a blueprint for us, just as for a car, only more complicated and hidden.

But these perspectives are, I maintain, not natural to humans—at least, not in their mature current forms. Rather, the mature intrinsic and extrinsic perspectives had to be constructed, sometimes laboriously, over millennia. It would take two discoveries, that of literature and that of science, to mature them fully. Humanity started off instead with merely a baseline view on the universe, consisting of whatever was useful to hunter-gatherers. It was the naïve perspective of primates, who cared only about using tools for personal advantage and about manipulating social hierarchies and that's about it. So it's a civilizational achievement to be able to extrinsically see the universe "from the outside." It is also a civilizational achievement to be able to intrinsically see the universe "from the inside." The two perspectives are the sources of our greatest triumphs, like our ability to observe galaxies light-years away, and also the elegance and beauty of the stories we tell. Although not technological marvels we can take a picture of, the intrinsic and extrinsic perspectives are *conceptual* marvels, and took as much intellectual work to create as our greatest institutions and constructions. They are, if judged by their fecundity, the cognitive Wonders of the World.

The history of these two perspectives, and their eventual entwinement in the form of an attempted science of consciousness, is the subject of this book. For it is in the modern science of consciousness that the two perspectives come to a head.[1-3] Of course, to tell this story in a single book (let alone deal justice to their fraught contemporary relationship in the science of consciousness) is an admittedly impossible task. And it is probably for good reason that I have never seen the story sketched in full as I attempt to do here, and therefore I must admit up front that there will be simplifications, perhaps even errors and oversights and omissions, all by necessity. To make it more manageable, when it comes to historical sources I will openly focus solely on what is sometimes called "the West"—with all the biases and limitations in scope that implies. But such focus reveals a clearly delineated arc. It starts with a dearth of the intrinsic perspective in ancient Egypt, then tracks its eventual development during Greek civilization, sees its full flowering in Roman times, and then follows the decline of the intrinsic perspective in the Dark Ages as literature once again ruminates little on interiority, until the intrinsic perspective returns as humans begin to purposefully surpass ancient wisdom, reaching its zenith during the Enlightenment in the development of literature and the novel.

Intriguingly, there is a comparable arc when it comes to the extrinsic perspective. Science begins off the board entirely, and early on in human history nature is filled with minds, purposes, teleology, and gods. This confused mix of the two perspectives lasts even through the height of Roman intellectual activity. What little extrinsic understanding of the world there was declined during the Dark Ages after the fall of Rome, before a resurrection of it during the Enlightenment.

It takes the scientist Galileo Galilei to fully understand the importance of cleaving the intrinsic from the extrinsic, allowing the extrinsic perspective to crystallize alone by itself into science. And yet, despite all the progress that has been made since then, it is increasingly clear that science cannot ignore the intrinsic perspective. Neuroscience and psychology run into invisible walls, and our understanding of the brain remains partial, at best. It is left to our generation to put the two perspectives back together in a science of consciousness.

Who am I to write this book with such a span that it involves not just history, but literature and neuroscience and philosophy and mathematics? It is impossible in scope. But if not me, then who? For I have lived for years ensconced in both perspectives, and feel, at a personal level, the tension in their paradoxical relationship. I grew up in my mother's bookstore and, later in life, became a novelist. Yet I am also a trained scientist. And in graduate school for neuroscience I worked on a small team advancing the leading scientific theory of consciousness. So for decades I have lived in the epistemological hybrid zone where the intrinsic and extrinsic perspectives meet. What I saw nearly blinded me with its beauty and paradox.

This book is an expression of what I've learned living in the hybrid zone. The early chapters, drawing on sources from ancient Egypt to Enlightenment salons, cover the history of the two perspectives, outlining how humanity forged them into what they are today. Then, jumping to the present, the book discusses the problems and perils that neuroscience faces, which stem from its purposeful ignorance of the intrinsic perspective, and how this oversight has led to a scientific crisis and a lack of progress that threatens neuroscience itself. Further chapters explain the growth of consciousness research as it attempts to reconcile the two perspectives. In turn, this brings to the fore questions of whether the two perspectives can indeed ever be reconciled, or whether science will remain necessarily incomplete, much as Kurt Gödel showed mathematics must. Drawing on my own research in neuroscience, causation, and information theory, the book concludes with how our evolving knowledge of the extrinsic perspective has revealed a scientific definition of "emergence" and therefore also free will—changing our very conception of ourselves.

But it begins, as all books do, with another book.

CHAPTER 2

The Development of the Intrinsic Perspective

In 1976, another researcher interested in consciousness, Julian Jaynes, then a professor at Princeton, published what would become a cult classic. The book was *The Origin of Consciousness in the Breakdown of the Bicameral Mind*. To this day it still sells well, and as a young man I read it and found it fascinating. For in it Jaynes proposes a radical theory: that prior to around 1200 BCE, humanity was a set of "automatons who knew not what they did," lacking all consciousness. Furthermore, he proposed that consciousness came about via a change in the relationship between the two hemispheres of the brain. Prior to this integration (which Jaynes says was driven culturally, but also by a modicum of natural selection), early consciousness had taken the form of things like auditory hallucinations, which were often interpreted as the commandments of the gods.

It's a wild theory. His evidence was only a careful textual analysis, particularly of Homer's epic the *Iliad*, which prompted Jaynes to note that the ancient Greeks often used the gods as stand-ins for discussing their own consciousness, such as when debating moral dilemmas or deciding how to act. Jaynes writes:

There is in general no consciousness in the *Iliad*. . . . And in general therefore, no words for consciousness or mental acts. The words in the *Iliad* that in a later age come to mean mental things have different meanings, all of them more concrete. The word *psyche*, which later means soul or conscious mind, is in most instances life-substances, such as blood or breath; a dying warrior bleeds out his psyche onto the ground or breathes it with his last gasp. . . . Now this is all very peculiar. If there is no subjective consciousness, no mind, soul, or will, in Iliadic men, what then initiates behavior?[1]

It's a keen observation of the paucity of description of people's inner mental lives in ancient literature. However, ever since the publication of *The Origin of Consciousness* its critics have consistently pointed out two things.

The first is that beyond the textual analysis the book contains no actual evidence that people were not conscious prior to the Homeric age, and therefore, this hypothesis of consciousness springing as full-formed as Athena did from the head of Zeus during the age of Homer is almost certainly untrue, especially given everything we know about the biology of the brain and evolution. While there have been some changes to the human genome in the past several thousand years, they have mainly been around pigmentation, nutritional intake, and body measurement.[2] And saying that consciousness is entirely cultural denies what many find unquestionable, like consciousness in animals or babies or toddlers who lack complex speech. So the idea that consciousness sprang onto the scene sometime in recorded history is too fantastical a claim to be backed up by only the flimsiest of evidence.

The second issue that's been consistently pointed out is that there is an eminently more sensible interpretation of the textual evidence Jaynes presents, which is that it is our *understanding* of consciousness that has evolved since ancient times. Consciousness was not absent in ancient Greece, but rather, at least broadly speaking, that is around the time in history when a more detailed understanding of consciousness began to develop and find its way into text. Here is the philosopher Ned Block in 1977, soon after the book's publication, writing a review of *The Origins of Consciousness* in the *Boston Globe*:

> But even supposing Jaynes is right about bicameral literature, there is a better explanation of this "data". . . . It is far more plausible to suppose that their basic processes of thought and notion were like ours, though they had a bizarre theory about these processes. Indeed, throughout the book, Jaynes confused the nature of people's thought processes with the nature of their theories of their thought processes.[3]

And this counterproposal that Jaynes's thesis applies to descriptions of mental states (rather than their existence per se) is still pointed out to this day, both in follow-up academic work and in reviews.[4-6] This weaker version of Jaynes's hypothesis, by benefit of being less extreme, is actually more interesting than the stronger version. Ultimately, it was the hewing of Jaynes to the stronger version that was the reason his theory never gained widespread scientific acceptance, being treated instead as a novelty or a lark by those who study consciousness scientifically, despite the elegance of Jaynes's writing and his marshaling of an impressive array of textual evidence. Sometimes those making this criticism of Jaynes's work have referred to what he called "con-

sciousness" as instead "theory of mind"—and while this description certainly applies, traditionally in science "theory of mind"
means predicting people's behavior based off of understanding
of their mental knowledge, such as when you know they'll behave in a certain way because they know or believe something.[7]
While our theory of mind certainly did develop over time—as
we will see, even the ancients had theory of mind—if you examine their literature, they were not "mind-blind"—rather they just
gave interiority short shrift. Supporting this, many scholars have
agreed that there are surprisingly few descriptions of mental
life in ancient literature, and that this is something that changes
over time[8]—a recognition that predates Jaynes's book.[9] It seems
that humans started knowledgeable about theory of mind but
without a well-developed intrinsic perspective on those minds,
and then slowly learned better how to represent, navigate, and
understand their own and other's interiority, both when it does
and doesn't relate to behavior. And the same is also true of the
intrinsic perspective's polar complement, the extrinsic perspective, what philosopher Thomas Nagel called "the view from nowhere"[10]—and what we now call science.

So, back to the beginning. Zoomed out so that all of history
is a mere panorama, its figures like children's toys from a great
height, its nations merely shifting lines on a map, its centuries
mere paragraphs, we will see that the development of civilization
is the story of humans exaggerating their baseline understanding
of the world in two opposite directions. Indeed, it is arguable that
what we even mean now by "civilization" is merely a society that
possesses fully developed versions of both perspectives, a society
that can switch between the two as it suits them, a move we each
individually make unnoticed with an almost aristocratic leisure.
But it was not always thus. We are the recipients of a gift we have
forgotten is a gift.

Ancient Egypt

It is difficult to overstate how different the ancient Egyptians were from us. Cleopatra herself lived closer in time to us than to those who built the pyramids of Giza. You can feel this distance for yourself—even just when viewing a sarcophagus at a museum one humbles at the alien iconography.

Of course, in other ways, they were just like us. A standard Egyptian home would have included a kitchen and a couch, and there were husbands and wives and children. There were shops and trades, fine linen and gorgeous pottery, schools, even a postal service. And when it came to makeup and ornamentation, which both the men and women wore, from eyeliner to body paint, the ancient Egyptians reached a peak we to this day have not surpassed in skill. Just like today, people were already complaining that culture was over, that there was nothing left to say, or do, and that all the greats were in the past.

> Would that I had words that are unknown, utterances and sayings in a new language, that hath not yet passed away, and without that which hath been said repeatedly—not an utterance that hath grown stale, what the ancestors have already said.

These words are more than four thousand years old, written by Khekheperre-Sonbu, a priest.[11] In many ways, civilization was there from the beginning, completely recognizable.

But in other ways, the ancient Egyptians were almost childlike. There was no concept of a vanishing point in Egyptian art; size indicated not nearness in perspective, but merely importance.[12] This shallowness included their conception of the human mind. They were, in a way, still carving off the human from the animal, both in their art and their gods (which were often half-animal, half-human). As Gottfried Richter writes of the Sphinx in *Art and Human Consciousness*:

> A human head is struggling to escape from the body of an animal. That was Egypt: still crouched down heavily in the animal's horizontal position, and completely devoted to the breath and pulse-beat of overwhelming cosmic forces with by far the greatest part of its being.[13]

This connection to "overwhelming cosmic forces" meant that, just as Jaynes points out, in Egyptian writing the inner workings of minds were often dramatized as a conversation with a god or spirit. Consider one of the most prominent classics of Egyptian wisdom literature (a genre called *Sebayt*), "Dispute Between a Man and His Ba," which is likely from the Middle Kingdom (2040–1782 BCE). A "ba" is the Egyptian notion of a soul/mind, or, at least, a ba is a part of the soul/mind, a part capable of traversing between the material and spiritual planes. In this way, the poem is very much about a man talking to himself about the inevitability and universality of death. But interestingly his mind speaks to him disembodied, entirely other.

> But behold! My ba would deceive me, but I heed him not,
> While I am impelled toward a death whose time has not
>   yet come.
> He flings me on the fire to torment me . . .
> My ba is senseless in disparaging the agony in life
> And impels me to death before my time.
> And yet the West will be pleasant for me, for there is no
>   sorrow there.
> Such is the course of life, and even trees must fall.

So trample down my illusions, for my distress is endless!
What my ba said to me:
"Are you not a man? At least you are alive!
So what do you gain by pondering on your life like the
    owner of a tomb,
One who speaks to him who passes by about his life on earth?
Indeed, you are just drifting; you are not in control of
    yourself."[14]

This tendency of associating the intrinsic with something extrinsic, like a god speaking, is identified by Julian Jaynes as evidence that consciousness itself was still only just coming to being, or remained fundamentally split—and he goes through the extensive textual evidence of this from ancient Mesopotamia to Egypt. But along the way Jaynes leans heavily on the idea that any evidence of consciousness in ancient literature is a result of mistranslation, as translators accidently import our own rich understanding of the intrinsic perspectives. He calls such changes to translations "modern mental impositions."

But Jaynes's theory, at least its strictest form, seems to break down in interesting ways once any credence at all is given to translations of ancient literature. For, despite this tendency to portray minds as extrinsic, if you look at ancient Egyptian writing, it comes across as very aware of what we might call a "social self." They understood how they came across to other people, especially if they were esteemed (or not) in the eyes of the greater group. We can observe this especially in Egyptian "autobiographies," which were already ornate and common by the Sixth Dynasty (2345–2181 BCE) and which were a way to highlight Egyptians' lives and legacies, their accomplishments, their bid for immortality and resurrection in the next life—in other words, they were their epitaphs. These autobiographies concerned mostly extrinsic events, people's actions, yet at the same time high-lighted an awareness of the perception of others. Here's the *Stela of the Chief Treasurer and Royal Chamberlain Tjetji* (Eleventh Dynasty, ~2070 BCE), one of the more renowned biographies owned by the British Museum (if just for its size alone). It tells of the life of Tjetji, the treasurer of a line of kings with whom he had gained great favor. In the stela images he is depicted surrounded by text, which reads:

> I am wealthy, I am great; I furnished myself from my own property, given me by the majesty of my lord, because of his great love for me. . . . Never did he find fault with me because my competence was great.[15]

There is prominent acknowledgment of the favor he garnered in his lord's mind, as the Egyptians were keenly aware of what others thought of their behavior, even, as demonstrated below, the people they governed. Like Sheshi, who was a vizier to Teti in the Sixth Dynasty at around 2300 BC, overseeing the documents and managing the office of the king. This inscription sits on the false door of his tomb (false doors were often used for autobiographies):

> I have come from my town,
> I have descended from my nome,
> I have done justice for its lord,
> I have satisfied him with what he loves.
> I spoke truly, I did right,
> I spoke fairly, I repeated fairly,
> I seized the right moment,
> So as to stand well with people.
> I judged between two so as to content them,

I rescued the weak from one stronger than he
As much as was in my power.
I gave bread to the hungry, clothes (to the naked),
I brought the boatless to land.
I buried him who had no son,
I made a boat for him who lacked one.
I respected my father, I pleased my mother.
I raised their children.[16]

There are plenty of similar examples, like *The Autobiography of Weni* and *The Autobiography of Harkhuf*.[17] It is unclear how these translations, which are often quite simple, could all be "modern mental impositions."

For while they are merely a laundry list of extrinsic accomplishments, such texts prove that the ancient Egyptians were not actually "mind-blind"—including about themselves. Clearly, the ancients possessed theory of mind and cared about pleasing their mothers and doing justice to their lord and standing well with people. This is unsurprising. After all, we primates evolved in an environment where gossip and other people's opinion, who was popular and who was not, were of great import.[18] However, an understanding that others have minds and opinions of us, of the social self, is not the same as having a well-developed intrinsic perspective. For the ancient Egyptians, characters' reactions, even their emotional ones, are flat and simplistic, playing out on the surface, close to behavior. It is as if they did not know how deep minds go. What they lacked was a good language for the subtleties of the mind, for its internal structure, what is called phenomenology.

There is a saying in philosophy of mind that "phenomenal consciousness overflows access consciousness"—terms introduced by none other than the philosopher whose original book review of *The Origin of Consciousness* pointed out that Jaynes conflated having a mind with the ability to describe minds: Ned Block.[19] The saying means that our experience (phenomenal consciousness, what it is like to be us) outstrips in complexity our ability to express it (the parts of our consciousness we can express). Using this terminology, we can see that the development of the intrinsic perspective is not the understanding that we and others possess consciousness—humans have known this since well into prehistory. And because of this, they have also been in charge of their own actions, not whipped around by gods as Jaynes would have it. But something did develop, which is our ability to access more and more of our phenomenal consciousness. That is, the development of the intrinsic perspective was the process of evolving our language and concepts such that the richness of access consciousness began to approach the richness of phenomenal consciousness, which is the situation we find ourselves in today. And this is what the ancient Egyptians and others of that time, at least if we judge by their literature, seemed to lack—access to the depths of their own consciousness.

There are actually early hints of the intrinsic perspective in ancient Egyptian poetry, although it is undeveloped, quite nascent. Some of the most evocative descriptions of minds in ancient Egypt are in love poems, some of which were written by women (or at least, from the perspective of a woman). Here, for example, is an excerpt from a set of love songs that trade off between a male and female speaker. That's a rather advanced literary technique, and it's notable that it was written a thousand years closer to us than any of the texts included so far (Dynasties 19 and 20, ~1292–1077 BCE). The poem, which belongs to the series *Beginning of the Songs of Delight*, is also housed at the British Museum. Here is one translation of this poem over a century old. It is the translation that first prompted me on this quest to understand ancient minds, because it was read at my wedding:

> And thou art to me as the garden
> Which I have planted with flowers
> And all sweet-smelling herbs.
> I directed a canal into it,
> That thou mightest dip thy hand into it
> When the north wind blows cool.
> The beautiful place where we take a walk,
> When thy hand rests within mine,
> With thoughtful mind and joyous heart
> Because we walk together.[20]

One can see the intrinsic here in a few ways, like the comparison of an individual to a garden (implying an internal topography), as well as the vividness of the accompanying sensory experiences, and the pleasure taken in the frisson of dipping a hand into cool water—not to mention the more obvious bits, like "thoughtful mind." It is impossible to read such a poem and believe, as Nietzsche did, that love was an invention of the troubadours.

Yet, despite the obvious evidence of the intrinsic here, Jaynes would likely raise a hand in protest. Is this just one of his "modern mental impositions"? For as he writes in *The Origin of Consciousness*: "Modern translators, for the sake of a supposed literary quality in their work, often use modern terms and subjective categories which are not true to the original."

That is, translators, for whom the intrinsic perspective is such an innate and even unnoticed way of viewing the world, synonymous with literature itself, find it impossible not to read into hieroglyphs emotions and mental states and poetics that aren't there. To see Jaynes's point, consider this more recent translation of the same poem by a different translator:

> I belong to you like this plot of ground
> That I planted with flowers
> And sweet-smelling herbs.
> Sweet is its stream,
> Dug by your hand,
> Refreshing in the northwind.
> A lovely place to wander in,
> Your hand in my hand.
> My body thrives, my heart exults
> At our walking together.[21]

Not quite as nice for a wedding. For this translation has far less of the intrinsic in it—it starts by comparing a person to a piece of land, a thing to be owned. The only real hint of the intrinsic is "my heart exults," and even here it is unclear to what degree the author conceptualized a heart as metaphorical of mind vs. body.

So is the more or less intrinsic one the better translation? Perhaps it is in the eye of the beholder. Jaynes was correct—translation is an art, and this clouds our judgment and probably biases us to reading more of the intrinsic perspective than was actually there (meaning that we should be skeptical with sources that radically depart from the norm).

Despite this unresolvable ambiguity, however, we can say this at least: there does seem to be mostly very shallow descriptions of mental life, very little of the intrinsic, in ancient Egyptian writing, at least compared to the modern day. What amount there is remains a subject of interpretation and translation and, at least in the writings we have, crops up in things like the romantic love poems that survived.

This is a nice thought, is it not? That we first began to understand how deep our minds go through love?

Ancient Greece

A preponderance of evidence for the lack of an intrinsic perspective in the early literature of ancient Greece can be found in *The Origin of Consciousness*. As Julian Jaynes writes, the *Iliad* is much like ancient Egyptian autobiographies, focused entirely on extrinsic actions and deeds:

> The characters of the *Iliad* do not sit down and think out what to do. They have no conscious minds such as we say we have, and certainly no introspections. . . . The *Iliad* is about action and it is full of action—constant action. It really is about Achilles' acts and their consequences, not about his mind.[22]

However, as Jaynes notes, the *Odyssey* is very different already from the *Iliad*—so much so that Jaynes suspects it was written hundreds of years later and by a different author.

> After the *Iliad*, the *Odyssey*. And anyone reading these poems freshly and consecutively sees what a gigantic vault in mentality it is! . . . It is a journey of deviousness. It is the very discovery of guile, its invention and celebration. It sings of indirections and disguises and subterfuges, transformations and recognitions, drugs and forgetfulness, of people in other people's place, of stories within stories, and men within men.[23]

As the centuries go by, there is more and more evidence of an increasingly well-developed intrinsic perspective. Particularly in ancient Greece and the lyric poets like Sappho and Pindar and Simonides. This development of the intrinsic perspective in lyric poetry even led to what can only be described as one of the first "intrinsic technologies" based on the techniques of Simonides of Ceos (556–468 BCE), whom Gotthold Lessing called "the Greek Voltaire" for his poetry.[24] Simonides's most impactful contribution to history was the invention of "the art of memory"—a mnemonic technique that relies entirely on mental imagery.

The story goes that Simonides had attended a banquet and, as a poet, gave mellifluous praise to his host, but also the twin gods Castor and Pollux, patron gods of sailors (and used to traveling to the mortal plane in the guise of St. Elmo's fire). In anger, his host told Simonides that he would only be paying half the agreed-upon fee, and wished him luck collecting the other half from the gods he had praised instead. Shortly thereafter, Simonides received a message. Apparently, two young men had requested he meet with them outside the banquet. The mysterious young men had already vanished, but as soon as Simonides left the building, an earthquake struck, and in tragedy the roof caved in over the great table. Only Simonides survived. The bodies of the guests were crushed, and none could be identified, so proper burial was impossible. But Simonides realized that by focusing on the spatial location of each guest seated around the table, he could remember them clearly.[25]

From this, Simonides began to make creative usage of the fact that the mind's eye can remember locations with ease. This art of memory begins with a "memory palace" that is generally a place you know well. And the specific things you want to remember, like the points you want to cover in a speech, are "placed" at various parts of the palace that you then mentally walk through as you give your speech. The ancient version of the teleprompter, it was used for centuries and across cultures as a tool for rhetoric, although after paper became cheap and books plentiful, the art mostly vanished, its usefulness lost. I myself learned the ancient art in college, when a professor of cognitive science taught it to me for the purpose of reciting a long poem in front of the class. To my surprise, I found that I learned the poem not only in a day, but also could recite it line for line backwards as well, by simply reversing my walk through my memory palace. To this day, an annual art-of-memory competition takes place in the United States based on speed memorizing decks of cards and long strings of numbers, and still uses techniques from Simonides.[26]

Evidence of "intrinsic technologies" speaks to a shift. Like so many things in ancient Greece around that time, the intrinsic perspective seemed to leap forward, especially around Athens and in the time of the sophists—intellectuals for hire who practiced and taught classes in subjects we would recognize as modern, like rhetoric and philosophy.

We can also see the development of the intrinsic perspective clearly in the literature of the age. Take Euripides, born around 480 BCE, greatest of the tragedians of the time, dramatist, intellectual, and reportedly exiled in his old age for crimes similar to those of Socrates (the difference between the writer and the philosopher can be summed up in their reactions to their sentence: Socrates choosing death, and Euripides, a coward, as all writers necessarily are, choosing exile). The signature of his style was the treating of mythic figures as if they were people with relatable thoughts and drives, which meant treating them as if they had minds like ours. His early play *Medea*, a classic of the Western canon, is a feast of the intrinsic. Medea sets out to punish her unfaithful husband, and the play is her unfolding emotional reaction to an original betrayal, a reaction that leads her on to violence.

Medea discusses the triumph of her anger over her conscious reason thusly: "I know what crimes I am about to commit, but my anger is stronger than my reason, anger which causes the greatest afflictions among men."[27] Bruno Snell, author of one of the original books on the history of mentality, wrote in 1946 in *The Discovery of the Mind* that Medea is "the first person in literature whose thinking and feeling are described in purely human terms, as the products of a human soul and nothing else."[28]

Did the flowering of the intrinsic perspective in the literature of ancient Athens merely reflect a maturity that had its origin in some other source? Or did the causation run from literature to life? Perhaps it was the early dramatists like Euripides who formally developed the intrinsic perspective first and then popularized it.

I cannot claim such knowledge for sure. But I do know that on the slopes of the Acropolis in Athens, you can sit where the intrinsic perspective may have leapt forward: on the stony seats of the Theatre of Dionysus where Euripides's plays were performed. In the quiet it feels like you are waiting for the masked actors to begin their performance in the open air. Occasionally, if you wait long enough, you might see one of the many stray dogs that now wander the Acropolis, and you can watch it pick its way down the path near the stage, scabbed and with a limp, acting out one more tragedy unnoticed by all but you and the great stage of pitiless stone.

Rome and the Dark Ages

If you read the literature of ancient Rome, it is inevitable you will be struck by how it feels entirely modern. When we examine the state of the intrinsic perspective in ancient Rome, we find it close to complete—at least, among the educated classes and elites where literacy rates were high. They speak, in their letters, much as we do. Here is Marcus Cicero (106–43 BCE), Roman consul, orator, and champion of the Roman republic over authoritarian rule (and eventually the victim of one such attempted king, Mark Antony). While exiled from Rome, Cicero wrote to his family, who were still there. In the letters he bemoans the disgraceful treatment of them in his absence:

> Don't suppose that I write longer letters to anyone else, unless someone has written at unusual length to me, whom I think myself bound to answer. For I have nothing to write about, and there is nothing at such a time as this that I find it more difficult to do. Moreover, to you and my dear Tulliola I cannot write without many tears. For I see you reduced to the greatest misery—the very people whom I desired to be ever enjoying the most complete happiness, a happiness which it was my bounden duty to secure, and which I should have secured if I had not been such a coward.[29]

Cicero also made use of the art of memory for purposes of rhetoric (as did many at the time). When he is discussing Simonides, it is obvious that Cicero has a clear understanding of distinctions within consciousness, such as how sight is more important for memory than the other senses:

> It has been sagaciously discerned by Simonides or else discovered by some other person, that the most complete pictures are formed in our minds of the things that have been conveyed to them and imprinted on them by the senses, but that the keenest of all our senses is the sense of

sight, and that consequently perceptions received by the ears or by reflexion can be most easily retained if they are also conveyed to our minds by the mediation of the eyes.[30]

We see the same attentiveness to the intrinsic in the writing of Plutarch (46–119 CE), famous for his *Parallel Lives*, which are among the first biographies of historical figures that have the purpose not just of recording and analyzing the extrinsic historical events, but also acting as studies of human nature. In a letter to a friend, Plutarch pontificates on techniques to help keep one's composure and achieve a tranquil mind:

Even as, on the contrary again, "conscience, since I am conscious of having done terrible things" like an ulcer in the flesh, leaves behind it in the soul regret which ever continues to wound and prick it. For the other pangs reason does away with, but regret is caused by reason itself, since the soul, together with its feeling of shame, is stung and chastised by itself. . . . That lament, "None is to blame for this but me myself," which is chanted over one's errors, coming as it does from within, makes the pain even heavier by reason of the disgrace one feels.[31]

Interestingly, the quotes in this letter by Plutarch are actually from Euripides's play *Orestes*. Plutarch is using the early dramatist's work directly as a means to discuss the viciousness of guilt. Such sophistication in letters is reflected in Roman literature itself. In Roman poetry alone there is a well-developed intrinsic perspective evinced by writers like Horace, Ovid, and Catullus. And again, it is especially in love poetry that we see a full mastery of the mind and its personal contradictions and peccadillos. The poet Catullus writes of his lover Lesbia that:

I hate and I love.
Why I do this, perhaps you ask.
I know not, but I feel it happening and I am tortured.[32]

When one reads them, one must feel the Romans were us. Or rather, we became them. If this appears an exaggeration, meditate on what came after. The fall of Rome was the end of a world-spanning empire that had lasted almost a thousand years. It is difficult to understate what the economic, religious, and intellectual fallout of something like that is; it would be as if the entirety of the United Nations declined and fractured apart into warring blocks.

The intrinsic perspective was hit hard by Rome's fall. If the early explosion of the perspective tracked the popularity and development of the theater in ancient Greece, then the decline of the intrinsic perspective tracked the quiet drawdown of the theater under the new Christian leaders during and after the fall of Rome. The early Christians thought of theater and paganism as irrevocably linked, and the content of fiction morally corrupting and distracting from religious devotion. From a certain point on, the official policy of the church recommended excommunication for those attending theater on holy days and sought to refuse actors the sacraments. Eventually one of the Christian emperors, Justinian I, closed the last Roman theaters in the sixth century.[33]

After the fall of Rome, what writing survived focused mainly on the ecclesiastical.[34] As Bryan Ward-Perkins writes in *The Fall of Rome and the End of Civilisation*:

Almost all the references we have to writing in post-Roman times are to formal documents, intended to last (like laws, treaties, charters, and tax registers), or to letters exchanged between members of the very highest ranks of society. . . . Most interesting of all is the almost complete disappearance of casual graffiti, of the kind so widely found in the Roman period.[35]

Such a change cannot be overstated; even in the small Roman town of Pompeii, so tragically well preserved in 79 CE, there are more than eleven thousand instances of graffiti.[36]

This lack of attention by medieval authors to mental states has been noted, with some scholars pointing out that characters in medieval texts often express their inner states, like emotions and thoughts, solely with direct speech and gestures, that is, their extrinsic behaviors—the mind was, to medieval literature, close to a material entity.[37] As one specialist remarked, medieval texts are full of characters "constantly planning, remembering, loving, fearing, but they somehow manage to do this without the author drawing attention to these mental states."[38]

Just as how in ancient Egypt what little intrinsic perspective existed was often reserved for romantic poetry, during the Dark Ages the intrinsic perspective became focused around, and sometimes even reserved for, religious experience. There are cases like "Dryhthelm's Vision" from around the eighth century, which describes a Scottish man's near-death experience of the afterlife (his unexpected Lazarus-like return to life surprised everyone, apparently, except his wife). After the vision he became a monk, but not before dividing his wealth between his wife, his sons, and the needy. The vision, very similar to Dante's much later *Divine Comedy*, is a tour of hell, purgatory, and heaven, and like Dante guided by Virgil, Dryhthelm was guided by an angel. The following passage describes not heaven, but a place where the souls almost good enough to enter wait, which Dryhthelm is drawn to:

As he led me through the middle of those happy inhabitants, I began to think that this might, perhaps, be the kingdom of heaven, of which I had often heard so much. He answered my thought, saying, "This is not the kingdom of heaven, as you imagine."

When he said this to me, I hated very much to return to my body, since I was delighted with the sweetness and beauty of the place I saw and with the company of those I saw in it. However, I dared not ask him any questions, but in the meantime I suddenly found myself alive among men and women.[39]

Outside of the church, there is some evidence of the intrinsic perspective in examples like *Beowulf*, although again minds are treated somewhat shallowly[40]—and, interestingly enough, its characters are pagans, not Christians. Additionally, the exact dating of *Beowulf* is unknown, as the first manuscript of it we have is actually near the end of the Dark Ages, dated to sometime around 1000 CE.

While during the Dark Ages the intrinsic perspective did not disappear back to pre-Grecian levels by any means (at least in the pockets of literacy that remained), there was far more variety in the perspective's degree of maturity. When compared to its blossoming in ancient Greece and Rome, the intrinsic perspective during the Dark Ages appears to have shrunk in scope, as our representations of minds, at least outside of religion, became shallower once more for hundreds of years.

The Rise of the Novel

The words of Anselmo struck Lotario with astonishment, unable as he was to conjecture the purport of such a lengthy preamble; and though he strove to imagine what desire it could be that so troubled his friend, his conjectures were all far from the truth, and to relieve the anxiety which this perplexity was causing him, he told him he was doing a flagrant injustice to their great friendship in seeking circuitous methods of confiding to him his most hidden thoughts, for he well knew he might reckon upon his counsel in diverting them, or his help in carrying them into effect.

So reads a passage from the 1605 novel *Don Quixote*,[41] by Miguel de Cervantes, in many ways the first recognizably contemporary novel. This scene is an early fit of postmodernism: in the book a man reads to a crowd (that includes Don Quixote and loyal Sancho) from an entirely different manuscript, a book within a book, this one concerning Anselmo and Lothario, two friends in Florence, Italy. The passage tracks Lothario's complex reaction to Anselmo's proposal that Lothario test Anselmo's wife's faithfulness through trying to seduce her (thus: "a Lothario").

Passages like these are why novels are the ultimate expression of the intrinsic perspective. In fact, tracking the flowering of the intrinsic perspective through the Renaissance and the Enlightenment is effectively the same as asking: When do novels as a genre reach maturity? And the answer is, I personally think, in the advent of "psychological realism," the heyday being between 1850 and 1950. It is exemplified in texts like *Middlemarch* by George Eliot, which is commonly used as an example of exploring the depths of interiority. The following passage is highlighted in *The Emergence of Mind*, edited by David Herman:

Celia thought privately, "Dorothea quite despises Sir James Chettam; I believe she would not accept him." Celia felt that this was a pity. She had never been deceived as to the object of the baronet's interest. Sometimes, indeed, she had reflected that Dodo would perhaps not make a husband happy who had not her way of looking at things; and stifled in the depths of her heart was the feeling that her sister was too religious for family comfort. Notions and scruples were like spilt needles, making one afraid of treading, or sitting down, or even eating.[42]

Of course, there is a long list of other familiar names: Jane Austen looms large here as an example of an early author who uses access to her characters' minds to create peaks of heightened drama. In Ian Watt's classic 1957 *The Rise of the Novel*,[43] he pegs the age of Defoe, Richardson, and Fielding as the "true" birth of the novel. Regardless of these subjective rankings, it is obviously a spectrum that dates all the way back to Cervantes and before. But despite it being a spectrum, there is notably an objective deepening in descriptions of minds over time. Writers themselves began to take explicit note of this aspect of literature. Like Virginia Woolf's literary manifesto, "Modern Fiction," which explicitly embraces this: "Let us record the atoms as they fall upon the mind in the order in which they fall, however disconnected and incoherent in appearance, which each sight or incident scores upon the consciousness."[44]

Perhaps the defining step that set the novel down its path toward describing consciousness as minutely as Woolf pursues was the medium's turn away from the theater. As we've seen, it was often through plays and the theater that the intrinsic perspective was originally expressed, or works meant to be read aloud. Even in ancient Greece and Rome, the purpose of literature was often performative. It always had an extrinsic aspect to it—the actual physical event itself, such as the onstage acting. A novel eschews all such extrinsic trappings. Not meant to be performed, it is merely words on the page, meant to be read in privacy— this, perhaps, made it inevitable it would become the purest expression of the intrinsic perspective.

Additionally, novels solve what philosophers call "the problem of other minds"—the problem that we can never know for sure what a person is thinking, or, from a metaphysical perspective, if they even have a mind at all! We must infer, we must guess, we must speculate. Novels, however, take place in an imaginary world where the problem of other minds does not exist, where mental states, like rage or ennui, can be referred to as directly as one does tables and chairs.[45] There's an entire academic field that highlights this, like Dorrit Cohn's *Transparent Minds*, published in 1978 (just a few years after *The Origins of Consciousness*), in which she emphasizes that this is "the singular power possessed by the novelist: creator of beings whose inner lives he can reveal at will."[46] Or as another scholar put it: "Novel reading is mind reading."[47]

No other medium can mimic this ability. Which actually provides a continued justification of the novel as an artform. Compare novels to the most popular medium of our age: film. Movies necessarily take an extrinsic perspective on the world. The author Tom Wolfe puts it thusly in his essay "My Three Stooges":

But when it comes to putting the viewer inside the head of a character . . . the movies have been stymied. In attempting to create an interior point of view, they have tried everything, from the use of a voice-over that speaks the character's thoughts, to subtitles that write them out, to the aside, in which the actor turns toward the camera in the midst of a scene and simply says what he's thinking. . . . But nothing works; nothing in the motion-picture arts can put you inside the head, the skin, the central nervous system of another human being the way a realistic novel can.[48]

Funnily enough, what often makes us think of a movie as artful, rather than mere entertainment, is when filmmakers try their hardest to portray minds in a deep "character-driven" way. Meaning that we judge movies as artful when they attempt to approximate the novel. An ironic fact, since they can never approach the novel's mastery of this.[49]

My bias here is obvious. I grew up in my mother's independent bookstore, and came of age among its shelves, and worked there as a teenager hawking fiction to customers, and so feel that the novel is something special, and has been relegated to being an undeservedly less popular artform. I even think it's arguable that this shift has changed our understanding of our own psychology. For instance, Freud was the best thing to ever happen to film and television. Of all the many ideas that Freud advanced, the most popular, even to this day, is the notion of psychological trauma being a central explainer of people's behavior. This idea permeates our culture, despite research showing that even extremely traumatic events, like living through horrific earthquakes and disasters, leads only to a minority of victims experiencing predictable negative psychological effects like PTSD, and also, that people's prior personality (to the event) has a strong effect on whether negative outcomes develop.[50] This doesn't mean trauma isn't real, but I think it's possible that trauma's popularity as an explanation of behavior came about because traumas are extrinsic events—they are things that can be filmed, they can be *seen*, which in turn means they can literally be shown on-screen as flashbacks. I find it no coincidence that the rise of trauma as an explanation of human behavior just so happens to correspond with the rise of our dominant narrative artform, the extrinsic medium of film, and its replacement of the intrinsic medium of the novel.

Film, of course, is an incredible and beautiful medium, but it trends toward characters being mere billiards set to and fro by external events. Only novels can describe the deep whirlpools of human consciousness, which is never fully reducible to response to an external event; it's a gyre that turns in each of us with its own weather and can render us ciphers to one another, except on the page.

The Full Arc

Perhaps the best example of how the intrinsic perspective evolved can be shown in a single example: that of the same tale told in the different stages of the intrinsic perspective's development across the millennia.

One of the oldest of such tales is "The Story of the Shipwrecked Sailor," which dates all the way back to the Middle Kingdom of ancient Egypt (2040–1782 BCE). The text is so old it holds the Guinness World Record for the oldest signature on papyrus, that of the scribe who copied it, Amenaa.[51] In the story, an official returning on a ship from an expedition encounters a storm and is washed up on an island, the sole survivor. There he meets a serpent that can tell the future, although, interestingly, nothing is told in the way of the official's emotional reaction at the sight of this fantastic beast:

> I uncovered my face and found that it was a snake that was coming. It was thirty cubits long. His beard, it was greater than two cubits long. His body was overlaid with gold. His eyebrows were real lapis lazuli. He was bent up in front. He opened his mouth to me while I was on my belly in his presence. He said to me, "Who brought you?

Who brought you, commoner, who brought you to this island in the sea whose sides are in the water?"[52]

Aspects of "The Story of the Shipwrecked Sailor"—like this run-in with a monster after landing on a strange island—turn up in Homer's *Odyssey*, when Odysseus and his crew set ashore on an island and stumble across the cave of a cyclops. Here the natural terror of the men at the sight of the beast is actually noted:

But when he had busily performed his tasks, then he rekindled the fire, and caught sight of us, and asked: 'Strangers, who are ye? Whence do ye sail over the watery ways? Is it on some business, or do ye wander at random over the sea, even as pirates, who wander, hazarding their lives and bringing evil to men of other lands?' So he spoke, and in our breasts our spirit was broken for terror of his deep voice and monstrous self; yet even so I made answer and spoke to him.[53]

Jumping all the way to the present day, the same encounter between a man and a beast is described in *Ulysses* by James Joyce, a heavily modernized retelling of the *Odyssey*. The unnamed narrator (who parallels Odysseus's "nobody" pseudonym) of the chapter called "Episode 12" ends up at a pub. This is sometimes called the "cyclops" part of *Ulysses*, with Barney Kiernan's pub acting as the cave, being guarded by a "monster" (a dog) alongside the "cyclops" (merely a gigantic man with an eyepatch), and the unnamed narrator becoming increasingly nervous around the two:

So we turned into Barney Kiernan's and there sure enough was the citizen ["cyclops"] up in the corner having a great confab with himself and that bloody mangy mongrel, Garryowen, and he waiting for what the sky would drop in the way of drink.

There he is, says I . . .

The bloody mongrel let a grouse out of him would give you the creeps. Be a corporal work of mercy if someone would take the life of that bloody dog. I'm told for a fact he ate a good part of the breeches off a constabulary man in Santry that came round one time with a blue paper about a licence.[54]

Perhaps nothing better sums up the historic development of the intrinsic perspective than this journey from an emotionless reaction to the most fantastical of beasts, all the way to a fantastical internal reaction to the mere sight of a common dog and its owner. Humans, by finding our depths, learned to dramatize our internal lives through literature, and in doing so we learned how to make the mundane extraordinary.

CHAPTER 3

The Development of the Extrinsic Perspective

How the extrinsic perspective arose, or at least how its final form, science, arose, is a story more commonly told. For our purposes, what matters is how it historically split from the intrinsic perspective (especially if we are to later take stock of the current scientific attempt to enfold the intrinsic perspective within the extrinsic one in the science of consciousness).

Besides, at least speaking broadly, the arc of the extrinsic perspective parallels that of the intrinsic: periods of monumental progress advance both perspectives simultaneously. Sometimes they are quite tightly bound: Einstein introduced his new geometrical perspective on space-time in 1905, while Picasso premiered cubism in 1907. Perhaps such synchronicity is merely correlation. Or, perhaps, it is causation—both Einstein and Picasso had been influenced by *Science and Hypothesis*, a 1902 book by mathematician Henri Poincaré.[1] (Einstein had read it with interest; Picasso seems to have had the book explained to him by a friend.)[2] Notably, *Science and Hypothesis* contained within it meditations on how best to not only understand, but *draw* the fourth dimension: for Einstein, the fourth dimension was time itself, represented mathematically, and for Picasso, the fourth dimension became the out-of-time layered points of view of cubism.[3]

Regardless of whether the two perspectives rise and fall together due to correlation or causation, we should review the story of the extrinsic perspective separately, at least briefly. In ancient Egypt we see very little of it. Instead, we see a society of magic, of gods who talk, of religious revelations and explanations. Medical problems were still thought to be caused mostly by evil spirits, and while pharmacological remedies existed (made of plants, minerals, etc.), they had to be prepared via specific rituals to be effective. Every organ was the domain of a specific god. In such a fantastical world of ghosts and spirits we see again the animal body, and the human head emerging only slowly from it. But there are hints of what's to come. The impressive engineering of Egyptian architecture required precise geometrical calculations, and there appears to have been mathematical knowledge, some of it surprisingly well developed, like having already had pegged π at about 3.17, which is only slightly off.[4]

In a sense, the environment itself forced ancient Egyptians to develop this knowledge of geometry—the Nile irrigated the land around it via its foreseeable rising and falling; in this, it is perhaps the most generous river in all of history. As historian Will Durant writes in *The Story of Civilization*:

The dependence of Egyptian life upon the fluctuations of the Nile led to careful records and calculations of the rise and recession of the river; surveyors and scribes were continually remeasuring the land whose boundaries had been obliterated by the inundation, and this measuring of the land was evidently the origin of *geo*-metry.[5]

The Greeks believed that it was the Egyptians who had invented mathematics. The historian Herodotus claimed that Pythagoras spent time in Egypt, where priests had introduced him to mathematics.[6] And in ancient Greece there are some early attempts to provide extrinsic accounts of the natural world, like when Thales proposed that everything could be ultimately reduced to water.[7]

However, in other ways the extrinsic perspective was underdeveloped in ancient Greece. In a fully extrinsic view of the world, there are no minds, nor do things really have mental characteristics. But according to Aristotle, the structure of the world was maintained by two competing purposes, essentially mentalistic urges, which were the tendency for bodies and elements to seek their natural place.[8] While Greek thinkers such as Democritus came closest to isolating the extrinsic perspective, proposing that everything is merely atoms and void,[9] Aristotelian teleological notions of purpose in nature would dominate in universities all the way until the 1700s.

It was only later that early scientists of the Renaissance and the Enlightenment began to move past the medieval conception of physics, abandoning its focus on Aristotelian teleology.[10] We all know this story, which takes place in those critical centuries in Europe when scientific thought unfolds and matures. It can be seen in the exceptional progress of the time: heliocentrism, calculus, gravity, the heart as a pump, Boyle's law of gases, the law of refraction, the spectrum of light, electricity, telescopes, industrial machinery, and empiricism—in many cases the literal engines of our world.

But why the Europeans? Why then? Why not the Romans? Why not the medieval Song Dynasty in China, with its banknotes, printing, and gunpowder? The answer is likely a historical anomaly, one of those small differences that makes all the difference. It was luck, in other words. For there was in Europe a virtual community, one built entirely over the mail, a community that lasted for centuries. This self-titled "Republic of Letters" was started by the early humanists, such as Erasmus, in the 1400s, and it acted as a moveable salon. Voltaire, one of the later central hubs in the network, described it thusly in his history *The Age of Louis XIV*:

> This was then the golden age of geometry. Mathematicians sent frequent challenges to one another, that is to say, problems to solve, much in the same manner as it is said the ancient kings of Egypt and Asia sent enigmas to be answered by one another. The problems proposed by these geometricians were of a much more difficult nature than the Egyptian enigmas, and yet none of them remained unanswered, either in Germany, England, Italy, or France. There never was a more universal correspondence kept between philosophers than at this period, and Leibniz contributed not a little to encourage it. A *republic of letters* [emphasis added] was insensibly established in Europe, in the midst of the most obstinate war, and the number of different religions; the arts and sciences, all of them thus received mutual assistance from each other, and the academies helped to form this republic.[11]

The Republic of Letters was spread out not just among artisans and engineers, but also among intellectuals who competed over patronage jobs granted by states, early universities, and powerful lords. And due to the political fracturing of Europe, there were many competitive patronage jobs.[12] This created a market for ideas, since those seeking patrons could cross borders to other jobs in different political landscapes. The Republic of Letters was particularly benefited by women *salonnières* who ran intellectual salons, organizing and breathing vitality into the decentralized network.[13] These women also often made their own contributions in the form of essays, monographs, and books, like the influential politics of *Letters from a Peruvian Woman* by *salonnière* Madame de Graffigny.[14] The history of science might have been quite different had perhaps Erasmus not lived, or had the political situation of Europe been less fractured, or had the practice of noblewomen hosting salons not caught on as an aristocratic virtue.

To this day science is a continuation of this Republic of Letters, endowed with its habits, its codes of conduct, its standards of evidence. It is worth remembering that scientific "papers" used to be called "letters." But the republic has been transformed now into a globe-spanning virtual community held together by patronage jobs at major universities.

Philosophers of science have long looked for some "secret sauce" that would make science a discipline that inarguably improves—perhaps, they reason, the "sauce" is its direct relationship to reality, or its basis in falsification of theories by empirical evidence. But are other fields not in "direct relation to reality"? And modern science regularly speculates about things beyond observation.[15] Even if scientists are the only ones who test their theories (a highly debatable proposition), there are all sorts of ways in which one can test; not only that, there are an infinite number of useless tests. So, while empirical evidence is held in high esteem, it is not science's sole motivator. Science is as much driven forward by its community standards of open participation, debate, hard-nosed skepticism, and a not entirely rational belief in the possibility of progress itself. Most important of all, it is driven by its tone, which is that of reasonable civilized discourse, almost as if science were always being conducted in a salon over dinner, and this tone is an inheritance from the aristocracy that birthed science.

When it comes to the evolution of the extrinsic perspective, there are plenty of leaps forward. Oft-cited is the one that occurred when Francis Bacon forever wedded the practice of science to testing and experimentation, but I think there is actually a more critical moment, which was when the intrinsic perspective was explicitly cleaved off from science, when consciousness was declared a no-go zone for scientific inquiry. It's even arguable that it's been science's adherence to giving only extrinsic explanations which has been the true root cause of its success. This cleaving of the intrinsic from the extrinsic was performed by none other than Galileo Galilei (1564–1642). In his witty scientific manifesto *The Assayer*, in 1623, he writes:

> Philosophy is written in this grand book, the universe, which stands continually open to our gaze. But the book cannot be understood unless one first learns to comprehend the language and read the letters in which it is composed. It is written in the language of mathematics, and its characters are triangles, circles, and other geometric figures without which it is humanly impossible to understand a single word of it; without these, one wanders about in a dark labyrinth.[16]

When championing that science should be in the language of mathematics, Galileo argued that science should focus on four properties of matter, all of which are susceptible to mathematical formalization: size, shape, location, and motion. By viewing the external world as solely a complex function of such extrinsic properties, Galileo was the first to fully formalize the extrinsic perspective, declaring the extrinsic the sole purview of science and suitable for describing the entire universe.[17] And ever since, science has focused on what can be measured and counted—and has, needless to say, had incredible success from doing so. But this means that, from the beginning, that success has been predicated on the removal of consciousness from the purview of science.

Of course, philosophers and thinkers noticed Galileo's bracketing aside of the intrinsic. Indeed, they even began to develop a specific terminology that attempted to isolate what exactly was missing from the extrinsic perspective. It was philosopher Charles Peirce in the 1800s who introduced the term "quale" to refer to how "there is a distinctive quale to every combination of sensation . . . a peculiar quale to every day and every week—a peculiar quale to my whole personal consciousness."[18] "Quale" comes from "quiddity," which means the inherent or ultimate essence of something—how a thing *is*. William James, one of the founders of psychology, and brother to Henry James, the novelist, made use of the same term. Even now, contemporary philosophers like David Chalmers use "qualia" as a synonym for what is intrinsic about "consciousness," or "experience," or "phenomenology," or "subjective experience," etc., where "qualia" refers to the what-it-is-likeness of a particular sensation or thought.[19]

In spite of Galileo's bracketing aside of conscious experience for hundreds of years, it was inevitable that modern science—specifically, the branch of the scientific tree we call neuroscience—would attempt to incorporate the intrinsic perspective. Attempt to explain the intrinsic in terms of the extrinsic. Attempt to explain qualia.

Funnily enough, this would have flummoxed Galileo. After all, despite his famous conflicts with the church over geocentricism, he was a religious man. He believed in a soul, and thought that science was the business of understanding the designs and plans of an almighty God. Which means he would likely think that the scientific attempts to explain "intrinsic-ness" (what we call "qualia") entirely through an extrinsic perspective would be efforts fundamentally misguided. As contemporary philosopher Philip Goff imagines:

> If Galileo traveled in time to the present day to hear that we are having difficulty giving a physical explanation of consciousness, he would most likely respond, "Of course you are; I designed physical science to deal with *quantities* not *qualities*."[20]

That is, Galileo might say that science works great as long as it doesn't directly study souls—such things are beyond science, the purview of culture and religion. Galileo might urge that the two perspectives should be kept separate, and never the twain shall meet.

Perhaps this imagined reaction by Galileo has more truth in it than we'd like to let on. So often we scientists overstate our case to the public. To tell the truth, all scientific fields look less impressive from the inside. Neuroscience is certainly no exception in this, but it also seems to have unique problems. If we are being honest, neuroscience has not progressed in the same manner as other closely related fields of biology, like the rapid rise of genomics, virology, or molecular biology. Perhaps we ask too much of modern neuroscience—it bears the burden of trying to reconcile two vastly different perspectives on the universe, ones that civilization has been refining in opposite directions over thousands of years. Following Galileo's prompting, we have pushed back all non-numeric qualities, all purpose and teleological reasoning, all discussion of the intrinsic, to the three-pound object that sits inside our skulls. And this has allowed the rest of science to proceed unhindered, all to undeniably monumental success. But the purview of neuroscience is not like other sciences. It is where the intrinsic and extrinsic meet. Ontology and epistemology begin to quake, begin to break down in their distinctions, begin to merge into something dimensionless and unnamable.

This is hinted at by how, on close examination, our study of brains faces increasingly insurmountable problems. I first became aware of these problems as a graduate student in neuroscience, where I was shocked to find that the field is, in secret, outside the eyes of the public and the other sciences, floundering in its attempts to understand the intrinsic, failing in its attempts to explain what we demand it to about minds and consciousness. Its status is, secretly, a scandal.

CHAPTER 4

Neuroscience in Need of a Revolution

In the early 1990s a team of Italian neuroscientists in Parma, Italy, stumbled their way into a supposed scientific breakthrough.[1] They did this by eating lunch. Across from them sat jealous and hungry macaques, monkeys who were in the middle of an experiment as test subjects. Recording electrodes had been snaked into the monkeys' brains, right into the premotor cortex, electrodes that could pick up the individual discharges of firing neurons; in such setups, the neuronal firing is often displayed on a screen or even plays out as sound. And when you listen in to the *pop-pop* talk of primate neurons, as I have, it becomes impossible not to try to interpret them. *Pop! Pop!* What was that? A complaint? A noticing? A shift in attention? An eye movement? The dark totemic eyes of the macaque ask you: What do you hope to discover here? This isn't—they say with a grin—going to go the way you think.

So you eat your lunch. And as they did, the researchers in Italy noticed something remarkable. At each bite: *Pop! Pop!* Picture the sandwich being lifted, followed by the explosion of live rounds that neurons make when played over a speaker. Yet the monkeys weren't moving—and the area of the brain being recorded from, the premotor cortex, was supposed to control merely their hand movements. Somehow the neurons in the movement-dedicated premotor cortex were responding to the hand motions not of the monkeys, but of the humans lifting up their forks to eat the food the monkeys coveted (for monkeys are the most covetous of all creatures).[2] Aha! Areas of the brain for controlling movement must have a secondary purpose, a functional time-share, wherein they also form the neural basis of understanding the actions of others performing the same kind of movement (even others of another species).

The discovery was dubbed "mirror neurons"—which simply refers to a behavior that was occasionally observed in some neurons—and, over the course of the idea's more than thirty-year life cycle, it has embodied the boom-and-bust narratology that drives neuroscience. A search of PubMed for "mirror neurons" produces thousands of results, peaking in 2013 and then falling off as the paper boom subsided. Neuroscientist and bestselling author V. S. Ramachandran promoted it heavily, writing on Edge.org in 2000:

> The discovery of mirror neurons in the frontal lobes of monkeys, and their potential relevance to human brain evolution—which I speculate on in this essay—is the single most important "unreported" (or at least, unpublicized) story of the decade. I predict that mirror neurons will do for psychology what DNA did for biology: they will provide a unifying framework and help explain a host of mental abilities that have hitherto remained mysterious and inaccessible to experiments.[3]

Soon after the 2000 *Edge* article, citations containing the phrase "mirror neurons" started to steeply climb. Ramachandran followed the article up with a triumphal TED talk titled *The Neu-*

rons That Shaped Civilization. But since those early days of hype the actual mirror neuron literature has descended into a morass of skepticism and vagueness. Indeed, the mirror neuron hypothesis has been walked back so far that there is almost nothing left to the story. They have explained nothing. It was always the case that only very few neurons showed mirrorlike activity, even when researchers hunted for them, and some were sensitive to highly specific observed behaviors while others responded to almost everything, and all throughout the many experiments there were a ton of confounding factors, like the extremely reactive and changeable nature of the neurons to things like angle of view, or if different amounts of reward were involved, and so on.[4-8] Also, the idea that mirror neurons explain something special or unique about humans never materialized—in fact, neurons performing mirroring functions aren't even unique to humans or primates; they've also been identified in rats[9,10] and birds.[11]

The hypothesis advanced by Ramachandran and others that autism is caused by a lack of mirror neurons[12] may seem an understandable leap, but it has always lacked evidence.[13] Brain region activity in autistic vs. non-autistic individuals just isn't that different when it comes to doing actions vs. watching them.[14] If individuals with severe autism, who have serious difficulty understanding the actions of others, still have functional mirror neurons like anyone else, shouldn't this falsify the theory? And if the functional time-sharing of mirror neurons allows humans to truly understand others' actions, how is it possible there are brain-damaged patients who are poor at performing particular actions (or even unable to), yet they can grasp the meaning of identical actions when others do them?[15] It's almost as if understanding and action are entirely separate.

So what do we have, at the end? What initially appeared as a straightforward story, a possible narrative handle on the Big

Question of what makes the human brain so special, ended up slipping away. The initial observation refused to be easily generalized into any sort of understanding or causality or control. In other words, it was merely an interesting correlation.

One might respond to all this dismissively, admitting that sure, hype is impossible to avoid in neuroscience, and it's bad when it happens, but maintaining that it's all self-correcting, all fine in the end, right? This Pollyannaish perspective might be true if it were the case that mirror neurons were widely recognized to have been disproven, or falsified, and everyone could move on. Instead, the proponents of the idea have simply complexified their accounts to the point of absurdity,[16] as if in their hypotheses about the brain they are constructing a Ptolemaic geocentric model of the solar system replete with epicycles.

Instead of getting corrected, the hypothesis lives on like the undead, still true enough to cite but not true enough to lead to any actual understanding about how the brain works. And during my time as a neuroscientist, I have seen versions of this story repeat over and over. An initial claim (Mirror neurons! Grid cells! Memory replay in sleep! Errors in decision-making come from microlapses in attention!) followed by a many-years-long backing-away process as the initial claims are found to be difficult to replicate, or, more commonly, the story becomes so complexified with caveats that the ideas become gaseous, amorphous—unkillable yet utterly useless. For how many years have neuroscientists and psychiatrists told the public that depression is caused by a chemical imbalance in serotonin levels? And yet there is no proven link, after decades of exhaustive research, between depression and these levels.[17] It was merely medieval humors, resurrected for the modern chemical age. Neuroscience is full of such zombie ideas. So much so they make up the bulk of it. Which is an indication there's something wrong with neuroscience itself.

A Building Suspicion

Sometime around 2015 I was talking with the head of the Allen Institute for Brain Science, one of the most respected organizations in neuroscience. Over beers he told me that it was a widely known secret among those high up in pharmaceutical research that most of the basic neuroscience coming out of academia was overstated. Companies didn't just have trouble with replicating the research literature; in many cases it was more as if the studies weren't even wrong—they were just kind of useless (like the idea of "mirror neurons"). And indeed, many pharmaceutical companies have pulled out of neuroscience since the early 2000s, removing what skin in the game they can. This is unusual, especially because since 1972, and the discovery of the very successful Prozac, companies such as Pfizer and GlaxoSmithKline have brought drugs to the market, including blockbusters like Zoloft and Paxil, and a set of other companies, like AstraZeneca, Bristol Myers Squibb, and Amgen, have also had commercial successes. Now, following a string in the 2000s of what can only be described as failures, AstraZeneca, Bristol Myers, GlaxoSmithKline, Pfizer, and Amgen have backed away from committing too many resources to neuroscientific drug discovery.[18-20] Which reflects on neuroscience as a whole: in 2019, when Amgen abandoned their drug programs in schizophrenia and Alzheimer's disease, the head of R&D admitted that their exit was due in part to the understanding of neurological diseases as being "fairly rudimentary,"[21] which is a consequence of the opaqueness around brain function in general.

This cutting of losses reflected a clear-eyed view of the lack of progress in neuroscience itself. Consider one such case: a famous 1996 study published in *Science* supposedly showing that people suffering from depression (or at least some of them) possess a particular version of the serotonin transporter gene called

5-HTTLPR.[22] Over the next decade this research led to hundreds of studies on the gene and its variants—all the way to discussion of modifying the application of antidepressants in order to account for an individual's 5-HTTLPR status. The 5-HTTLPR genes were dubbed "orchid genes" and, perhaps inevitably, covered with breathless excitement in major media organizations like *The Atlantic* and on NPR.[23,24] They were so named, at least according to the excited researchers and journalists, because variation in the genes created individuals as delicate as orchids, plants which are beautiful but notoriously difficult to grow, and very susceptible to changes in environment. You might be able to guess the story by now. Once genetic analysis improved and as much larger sample sizes became available, follow-up research cast serious doubt on whether variations in 5-HTTLPR had any effect on depression at all.[25,26] How could hundreds of studies—all investigating a gene's relationship to mental illness, conducted over decades, and most of which purportedly showed a positive effect—turn out to be wrong?

It's been well known that there are growing issues[27] around replication in the biomedical sciences, particularly around clinical outcomes. Somewhere around half of highly cited biomedical studies, all of which supposedly had effective interventions, flat-out just didn't replicate when different teams of scientists conducted them independently. Companies such as Amgen and Bayer, those with money on the line, had themselves performed replication attempts revealing that a shockingly low number of papers (as low as 11 percent) within cancer biology had replicable effects.[28] Notably, even in cases where the effect did indeed replicate, the effect size (the degree to which the medical intervention actually had an effect) usually significantly decreased.[29]

The academic field of psychology has itself been suffering a similar crisis—over the past decade, large-scale replication studies have been conducted by multiple labs and, at best, have a 50 percent replication rate, and almost always the effect size declines.[30,31] Tenets of psychology, basic topics you can still crack open a psychology textbook and find results for, like "ego depletion"[32] or how powerful money is for behavioral priming[33] or the Stanford Prison Experiment,[34] and so on, arguably either can't be replicated or have only small effects once the data is analyzed appropriately.[35–37] Even the supposed Dunning-Kruger Effect,[38] in which people with low knowledge or experience in a given area supposedly overestimate wildly their expertise, may be a statistical artifact.[39,40]

This should immensely concern neuroscientists. After all, neuroscience is just a more complex form of psychology that attempts to ground psychology in the mechanics and dynamics of a hyper-complicated system, the brain—an organ that remains stubbornly hidden behind a wall of bone, and an issue regular psychology doesn't have to deal with. It's for this reason that neuroscience relies heavily on neuroimaging, but this merely adds more sources of statistical variability and opaqueness on top of those already in psychology. Most neuroimaging conducted in humans doesn't even analyze neural activity directly, but is multiple steps removed (fMRI tracks blood oxygenation levels of parts of the brain, not actual neuronal firing), and when it can track neural activity directly, as through electrophysiology, it does so only for an astronomically small percent of neurons. This poses problems for interpreting the results, not to mention all sorts of issues around effect sizes and sample sizes.[41] And yet there is a credence granted to reported results that seems impossible to penetrate. In a notorious study in 2009,[42] a dead salmon was put in an fMRI scanner and shown the kind of standard fMRI task of "looking" at photographs that depicted humans in social situations. The dead salmon, quite obligingly, showed a statistically significant response to a common analysis pipeline.

Far more fundamentally, and little discussed, is that almost all the reported effects in neuroscience are computed based on averages of brain activity. Why? Because the data from a single trial of a neuroimaging experiment is essentially always a complete mess. Because of this, the background assumption behind the majority of neuroscientific research is that the constant noisiness of the brain can be smoothed out by averaging. The assumption goes that while brain responses to individual stimuli look mostly just like noise, via averaging across a bunch of responses, the "real" structure of neural activity will pop out to experimenters. And such averages are pretty much what's reported and discussed in most neuroscience papers. But there's a problem: the brain itself doesn't appear to care about averages! Averages of brain activity are not differences that make a difference. Rather, they are abstract theoretical concepts compiled by scientists for papers. They are not used by the brain to *do* anything. These averages are, ultimately, just artifacts created by scientists. And there is growing recognition this is a problem. In 2021, researchers identified two criteria that, if the brain did care about averages, should be true:

> Reliability: Neuronal responses repeat consistently enough across single stimulus instances that the average response template they relate to remains recognizable to downstream regions.

> Behavioral relevance: If a single-trial response is more similar to the average template, this should make it easier for the animal to identify the correct stimulus or action.[43]

That is, they ask whether (a) individual neural responses can be reliably identified as belonging to one particular average response, and also if (b) the closer the match between the individual trial and the average, the better the animal's performance on the task is. Lo and behold, when the researchers tested this, it turned out that neither criterion strongly held true. When they started correlating electrophysiological recordings in mice brains to mice behavior, they found both that the single-trial response to a particular stimulus was difficult to identify with the average it was contributing to, and also that how well the brain's individual response matched the average had little, if any, bearing on behavior. That's extremely bad. It indicates that the statistical constructs used to create neuroscience papers are epiphenomena for the brain itself—as if we are trying to understand a clock by examining the shadows it casts.

The deep rot from examining mostly epiphenomenal artifacts crops up in the failure to translate the correlations of neuroscience into good predictions. For instance, we should be able to predict behavioral phenotypes (types or classes of human behavior) based off of the neuroimaging that is so commonly used for papers. We can even lower the bar to just throwing machine learning at the neuroimaging data and seeing if it can distinguish various behavioral phenotypes using any and all available correlations. A specific example might include telling the difference between a depressed person and a non-depressed person from merely neuroimaging data alone. Yet, even with extraordinarily large sample sizes with which to make such distinctions, like having a thousand participants, it turns out that using neuroimaging is not very accurate.[44] And this is a really big issue because neuroscience, as a discipline, has proceeded in an entirely "mom and pop" fashion. This means that professors have a half-dozen people in their lab, get access to an fMRI machine, and then book a couple dozen scans of undergraduates for whatever paper they're doing next. This is true for the majority of neuroimaging and therefore composes the bulk of the literature of neuroscience itself (or at least, the aspects of it focused on human cognition rather than, say, the molecular machinery of individual neurons). Because of this, the median neuroimaging study has a sample size of something like twenty-five participants. And there are really good reasons to think that's tens of thousands off—if not a couple million off—in terms of being a sample size large enough to make actual replicable claims. I railed against this in graduate school (much to the annoyance of my professors), and in the intervening years this issue has been supported by research showing that it takes thousands of individuals to achieve reproducible brain-wide associations.[45] This is not a bar most neuroimaging studies pass. But what's especially scary is that the field continues apace anyway, and there is little talk (compared to other fields like psychology) about the neuroimaging replication bubble.

Because of such problems, current neuroscience does not translate very well to actual knowledge about how the brain works. This can be seen simply by asking a neuroscientist to explain any cognitive function. Take note of the answer they give. Then ask them to explain again, but this time without any reference to spatial location. The vast majority of the time they will be flummoxed. "Speech is processed in Wernicke's . . . no wait." We know, very roughly, how some of the earliest sensory processing seems to work, like how edge detection occurs in vision. The shallows of the brain. All the deeper areas remain a mystery, beyond that some areas seem to be, statistically, more associated with some things than others (and thus "localized," although the degree of this localization is up for debate). But what does location tell us? What does mere *involvement* tell us, beyond what areas to avoid during neurosurgery? Almost nothing.

It gets worse. For the brain is plastic, extremely so. In the macaque visual cortex, about 7 percent of synaptic boutons are turned over every week[46] (boutons are the point of close contact between axons and dendrites, which is where communication between two neurons occurs). And that's in an area with a supposedly stable architecture that only does initial visual processing, which should be long finished in adulthood, like the detection of edges and contours. But this incredible degree of plasticity even in adults makes many of the traditional goals of neuroscience nonsensical, like tracking down "representations" in the cortex: identifying neurons that selectively fire only for certain concepts, such as only for pictures of your grandmother.[47] More recent research has shown that there is "representational drift" wherein all those supposedly important locations that neuroimaging is picking out, from brain regions to what individual neurons are selective for, aren't stable—rather, they drift and change across time, and will vary not just individual to individual, but even day to day within an organism. This occurs both at the complex higher-level areas of the cortex and also at the lower-level, supposedly more "stable" architectures like the primary visual cortex.[48] That is, the location of where a particular function is performed actually drifts across the brain like a slow-moving cloud. And this isn't some unique biological oddity—researchers have noted that this looks very similar to the drift seen in artificial neural networks, where the functions individual artificial neurons perform shift as the network learns.[49] The territory is changing faster than we can make maps.

Because of issues like this, the tough truth is that most neuroscientists spend their careers just splashing around in the statistical mud. Once the neuroimaging replication bubble inevitably pops sometime in the next decade or two, neuroscience should be very careful. There is more wrong with it than just that findings don't replicate, that the effect sizes are tiny, that every lab uses a different methodology, that small changes in methodology lead to big changes in outcomes, or that neuroscientists reinter-pret results as they go in a phenomenon that's been referred to as a "Garden of Forking Paths"—wherein choosing among the end-less permutations along the way to finishing an experiment all but ensures a statistically significant outcome.[50] There is a tendency for scientific fields under duress to focus on technical method-ologies, to drill down on issues like pre-registration of studies, identifying clear cases of "p-hacking," trying to prevent results from getting cherry-picked, trying to make studies have larger sample sizes, and trying to curtail scientists making up hypoth-eses to fit the data they've already collected.[51] These are all good things to correct, but often when a science doesn't know what to do or where to go, it becomes obsessed with drilling down on the methodology: as if science were simply an algorithm, like furni-ture assembly, a process that once purged to be rigorous enough ensures that whatever is cranked out of the algorithm is automat-ically a successful science. This isn't true. In science, just as in the rest of human endeavor, if you aim for a nonexistent target, your arrow will never hit anything at all.

And what is the target of the arrow of modern neuroscience? Uncertainty over this question reveals itself in the shifting termi-nology that neuroscientists regularly employ. The jargon changes amorphously from subfield to subfield, often based on the exact same empirical phenomena. If a brain area responds to outside stimuli via some change in neural firing, depending on the sub-field that response might be called "computation" or "represen-tation" or "storage" or "retrieval" or "replay" or "processing" or "information transmission." And the chosen term also depends on what narrative the authors want to tell; only rarely does it signify any fundamental difference in the neural activity being observed. Most of the time, all a neuroscientist can really say is that there was a change in neural activity. Therefore, when scientists try to understand the Big Questions of neuroscience, such as "How does the brain process information?," they run into irresolvable problems, since the methods of neuroscience only actually track what they seek once the nomenclature has done its work.

Some may reply that the failures of traditional neuroscience are simply a function of the immaturity of our ability to intervene on and observe the workings of the brain. Assuredly as technology progresses, and our ability to model and map the brain's connections improves, these are all just temporary challenges.

Unfortunately, no. For these issues go beyond the mere paucity of our tools. In the 2017 paper "Could a Neuroscientist Understand a Microprocessor?"[52] researchers used a model system to test neuroscience's ability to understand systems far less complicated than the brain itself. They took advantage of how the MOS 6502 microchip—which was used in the 1980s to power Nintendos—had been perfectly reconstructed (as a computer model) by retro-computing enthusiasts. Despite having only 3,510 transistors, the efficient simulated microprocessor could still run a few games: Space Invaders, Donkey Kong, and Pitfall.

Pretending to not understand the function of the chip, the researchers attempted all the common techniques of neuroscience: looking at the connectomics (or the wiring diagram of the chip); performing "lesions" on the transistors of the chip mimicking the ablation studies of neuroscience; analyzing the individual transistor behavior using the same techniques neurons are normally analyzed with (things like firing rate or tuning curves); correlating the transistors to each other in pairs; averaging the chip's activity into fMRI-like voxels over a small local region; along with other more complicated mathematical techniques that neuroscientists commonly use, esoteric analyses like dimension reduction and Granger causality. Yet despite knowing the complete wiring diagram, despite having all the microscale information that we *wish* we knew for the brain, all the conclusions the researchers could derive from the suite of techniques neuroscience has to offer were trivial or, in some cases, directly misleading. E.g., during the iterative "lesioning," the removal of roughly 50 percent of the transistors (1,565) had zero impact, while the other 50 percent (1,560) flat-out broke the chip. A tiny minority of lesions were game-specific, like the finding that 98 transistors, if lesioned, led to the unique failure of Donkey Kong to boot up (they had no impact on the other games). Aha! Perhaps these were the transistors that represented or computed Donkey Kong? It turns out, no. The reasons that Donkey Kong failed under these "lesions" had nothing to do with Donkey Kong as a game—what it represented, its function, or its information processing—but were the result instead of what was essentially a fluke involving the back-end code. And the other tools in the neuroscience toolkit led to similarly opaque results, things which, if taken seriously, would have led a researcher down rabbit holes and into confusing cul-de-sacs. Arrows were being fired at nonexistent targets.

Neuroscientists often assume that some simple, discoverable algorithm outputs behavior—some set of boxes and arrows that can be visualized and understood, that describe like an engineering diagram how some part of the brain works. But why even have this assumption? We know from research into artificial neural networks (ANNs) that as they grow in size they increasingly become a "black box" wherein there is no engineering diagram of what-does-what that makes sense, other than the system as a whole, which is impenetrably large. When it comes to an interacting web of billions of parameters, asking questions like "What does what?" or "Where is this information being processed?" or "What does this node contribute to?" often begin to stop making sense.[53] No one knows how the large-parameter models that show early signs of general intelligence, like GPT-4 or Google's

PaLM, actually work. We just know that they do. And this is because there is often no compressible algorithm that an ANN is implementing.

Applying this same reasoning to neuroscience leads to some uncomfortable conclusions. Neuroscientists often assiduously avoid such discussions, since asking "How does the brain perform this transformation between input and output?" is a far more complex version than that same question put to ANNs, and with ANNs we know that often in principle we can say very little about this (and that's with the complete and perfect access to the connectome, or wiring diagram, of the ANN, unlike the brain, which comes to us piecemeal via invasive surgeries or coarse-grained neuroimaging). So it's not a lack of data about the brain that's the problem. It's the approach.

But how can this be? For there is a great deal of neuroscience-ing going on; scientists all over the globe are getting their degrees and putting people in brain scanners and running MATLAB programs. There are so many new papers published every year, all those prestigious *Nature* and *Science* appearances, yet so little of it translates into fundamental understanding.

One must ask: If arrows are being fired at nonexistent targets, then what's the right target?

Neuroscience's Missing Target

There are two fundamental views of science: the incremental and the revolutionary. The great analyzers of science can be broken into the incrementalists, like Karl Popper, who thought science proceeded by the slow grind of falsificationism,[54] and the revolutionaries, like Thomas Kuhn, who thought that paradigm shifts swept away old sciences and made their previous results incommensurate with the new paradigm.[55] Then there are the moderates arguing for something in the middle, like Imre Lakatos, a Hungarian-born philosopher of science of the twentieth century.[56]

The Lakatosian view of science conceptualizes scientific theories as having a set of core principles protected by an outside belt of changeable principles, so that when a prediction is falsified, the changeable belt adapts. It is only in times of radical change that the core principles are overturned and some new set takes their place. This prevents a prevailing view from being overly fragile, wherein a single instance of any sort of falsification can overturn an entire paradigm. For this reason the Lakatosian view of science has been described as "refined falsificationism," essentially, a middle way between Kuhn and Popper.

Perhaps Imre Lakatos endorsed a middle way because in his life he had traveled the full scope of opinion from hardcore young revolutionary to older political skeptic—he had learned the value of tempering one's opinions. When young, he was a Communist helping to topple the Hungarian government (even convincing a young girl to commit suicide to protect Communist secrets), an informant, and an anointed member of the post-revolution government's thought police. It's possible that exposure to Karl Popper's work in midlife actually changed his attitude toward Communism, since he may have suspected that Marxism's historical failures in the form of gulags, starvation, and unsuccessful utopias effectively "falsified" the theory.[57] Having fled Hungary, and when older and living in Britain, Lakatos took a stand against the excesses of Communism, and in a way, his former self. At what point did his ideological view of the world eventually reach irreconcilability with what was going on around him and that last final falsification occur?

I confess to being mostly a Kuhnian, with Lakatosian sympathies about how scientists defending and working on theories actually behave. This means I think that, at least in general, progression within the Republic of Letters that we now call "science" means moving from a pre paradigmatic field to a post-paradigmatic one (or perhaps, if you're a Lakatosian, shifting on the spectrum from pre- to post-paradigmatic). Post-paradigmatic sciences are marked by their lack of weirdness—chemistry is, I am assured by the complacent way its graduate students act, quite normal, a veritable suburbia of science where experiments are done without philosophical or existential debate. Not so with neuroscience.

A field moves from pre-paradigmatic to post-paradigmatic mostly by the introduction of some new Big Theory. There is a reason we call it Einstein's *annus mirabilis*: the year of miracles. Darwin's theory of evolution in *On the Origin of Species* or Rutherford's model of the atom are such Big Theories. A field is pre-paradigmatic when it is fundamentally in crisis, a crisis stemming from some deep and unexplained question that influences everything but is ignored. Thomas Kuhn himself gives an elegant definition of these crises in pre-paradigmatic fields, awaiting the emergence of the new, in *The Structure of Scientific Revolutions*:

> If awareness of anomaly plays a role in the emergence of new sorts of phenomena, it should surprise no one that a similar but more profound awareness is prerequisite to all acceptable changes in theory. On this point historical evidence is, I think, entirely unequivocal. The state of Ptolemaic astronomy was a scandal before Copernicus's announcement. Galileo's contributions to the study of motion depended closely upon difficulties discovered in Artistotle's theory by scholastic critics. Newton's new theory of light and color originated in the discovery that none of the existing pre-paradigms theories would account for

the length of the spectrum, and the wave theory that replaced Newton's was announced in the midst of growing
concern about anomalies in the relation of diffraction and
polarization effects to Newton's theory. Thermodynamics
was born from the collision of two existing nineteenth-
century physical theories, and quantum mechanics from
a variety of difficulties surrounding block-body radiation,
specific heats, and the photoelectric effect. Furthermore,
in all these cases except that of Newton the awareness of
anomaly had lasted so long and penetrated so deep that
one can appropriately describe the fields affected by it as
in a state of growing crisis. Because it demands large-scale
paradigm destruction and major shifts in the problems
and techniques of normal science, the emergence of new
theories is generally preceded by a period of pronounced
professional insecurity. . . . Failure of existing rules is the
prelude to a search for new ones.[58]

Within neuroscience the anomaly, the failure of existing rules
that signifies the necessity of a future paradigm shift, is the difference between conscious and unconscious neural activity. There
is no boundary or border that can be drawn on any map of the
cortex that can currently tell you when neural activity crosses from
the extrinsic into the intrinsic. It is for this reason that neuroscience resembles so much the treading of water. It has not had its Big
Theory moment. The reason neuroscience is a quagmire is that it
ignores the brain's entire evolved purpose, it's very raison d'être—
maintaining a stream of consciousness. Every region or module
operates on and requires the stream of consciousness in order to
make sensible decisions—consciousness is like a brain-wide frame
of reference for every other cognitive function. It is the elephant in
the brain that neuroscientists feel only parts of, blindly.

Consciousness Is What Brains Are For

Instead of confronting the problem of identifying how consciousness works head-on, modern neuroscience performs a move that even Galileo would probably view as a perversion: pretending the problem doesn't exist. To make minds manageable, neuroscience has minimized the importance of consciousness—perhaps because, regardless of if new paradigms are needed, science as an academic system simply cannot wait for them. Even if it cannot make progress, science is forced into the appearance of progress by its institutions.

For this reason, neuroscientists, cognitive scientists, psychologists, and philosophers have all tried to downplay the anomaly that consciousness poses for neuroscience by suggesting that consciousness is some sort of minimal subsystem of the brain, possessing no information, almost useless. The steam from an engine. A hat on top of a hat.

But the importance of consciousness for neuroscience can be ignored only for so long. Consider the phenomenon of "change blindness," which is often portrayed in psychology textbooks as meaning that you are "blind" (or unconscious) to most of the world except a sliver of attention. Classic examples include experiments wherein participants' saccades (the micro-movements of their eyes) are being tracked while reading on a screen. Meanwhile, letters or words far from where their fovea (the high-resolution center spot of your vision) is tracking are minutely changed. The participants don't notice small changes to lettering, but it's well-established that participants do notice larger changes—it's just that those noticings aren't popularized, since it's not very interesting to publish a paper saying "Yes, people have peripheral vision and really do see the scenes in front of them," while it's very interesting to claim that we don't see what we think we do. Surely you are aware of the room around you in which you are reading this—if something were to move suddenly in your peripheral vision, you'd see it. You are aware of it, just in a more coarse-grained fashion than what's in focus. And while many cognitive scientists have argued that we do not actually experience color vision in the periphery of our visual experiences, this is just another neuro-myth—with the right experiments, people are able to differentiate colors at the very periphery of their visual field.[59] The same is true for identifying objects in the periphery, indicating a relatively fine-grained representation of space across the visual field, although crowded sets of complicated objects are often "chunked" in the periphery—again, exactly the conclusion that you'd arrive at if you introspected about your own conscious experience.[60] Our very survival as an organism depends on our stream of consciousness being constantly veridical and richly informative. And there are plenty of examples of this, such as when scientists asked individuals to search for a particular face within crowds, and even those faces that were non-targets and passed over quickly could be mostly recalled later, indicating that humans really do store even the minutest details of our stream of consciousness, at least in the short term.[61]

It is, of course, absurd that the state of neuroscience is such that I feel the need to tell people that they're actually experiencing what they think they're experiencing. But that's indeed the state of it. Just as you assume in day-to-day life, consciousness is actually immensely informative, and your phenomenology, the structure of your conscious experience, is rich and intricate and almost everything in the brain makes use of it.[62] You can just reflect on your visual awareness of the space around you, the phenomenology of your visual field. At every point it is occupied, usually with something different, and every point bears some relationship to other points. Even when looking at a black canvas your spatial experiences are "occupied" with the pointillism of perception.[63]

The informativeness of consciousness can actually be measured, or at least estimated in terms of how large it is actually is. As my mentor in graduate school, Giulio Tononi, has pointed out, you are currently occupying just one specific conscious experience, and this rules out all the other possible conscious experiences you could have had.[64] That is, you are seeing this text now, rather than watching a movie, and even every frame from every movie is just some tiny subset of the many differentiated conscious experiences you could have. And information comes, as mathematician Claude Shannon, considered the father of information theory, discovered, from the ruling out of alternatives. A coin flip has one bit of information because it rules out one alternative, meaning you can answer one yes-or-no question about it. How many yes-or-no questions could you answer about a particular conscious experience? The answer is extremely large.

If we acknowledge primacy of consciousness for the brain's function, what would it mean for neuroscience to become a post-paradigmatic discipline? We must ask, in a manner constructed from scratch: What does it really mean to understand the brain?

An analogy. When we look out at the physical world, we see at first a bewildering tapestry of movement, from the falling of leaves to the running water of a sink. But underneath, we now know, there are lawful actions behind each and every varied and diverse phenomenon. Therefore, the best understanding of the physical world is not one of statistical correlations but rather the regularities we call the "Laws of Physics." And despite the varied appearance and expressions in the rich phenomenology of nature, everything everywhere is presumably ticking away in lawful followings. At every second, every nanosecond, of time, the world unfolds according to these hidden laws.

So too it must be with the brain, wherein there is something lawful governing the flexing and shifting web of neural activity. There, the high-level laws governing neural dynamics are those of consciousness.

This is not to suggest that such laws supplant or contradict the underlying physical laws of the brain, any more than the Law of Least Motion at the classical scale supplants or contradicts the laws of quantum mechanics. In the brain there is, likely at some spatiotemporal mesoscale (higher than neurons, lower than entire brain regions), a set of lawful followings based on consciousness, and these are the hidden laws behind the varied appearance of neural dynamics. The following of these laws creates a structure we call "experience." The standard correlational approach of neuroscience is therefore mostly irrelevant. It is like trying to understand the physical by tracking only correlations—such correlations are present, yes, infinite, yes, but also useless without theory. The gold standard of science remains the lawful, meaning that a post-paradigmatic view of the brain should not be statistical, but lawful. Or to boil it down—at the highest level, and in terms of basic science, there are essentially just three neuroscientific discoveries we care about: (a) the biological mechanisms of the brain that underlie cognition (such as types of neurons, connectomics, the molecular study of synapses); (b) the learning rules of the brain, which specify how its high-parameter space of neuronal connections evolves; and (c) how activity in the high-parameter space lawfully gives rise to certain conscious experiences. Of these, (c) is the most ignored aspect of neuroscience today.

Just as how geneticist Theodosius Dobzhansky pointed out that "nothing in biology makes sense except in the light of evolution,"[65] it will eventually become clear that nothing in the brain makes sense except in the light of consciousness.

CHAPTER 6

The Two Houses of Consciousness Research

Why was consciousness ignored for so long? Just as in chess, the movement of a scientific field from the margins of the board to the center signifies major changes, and research on consciousness has inched closer and closer to centrality over the last few decades. Yet it started out not just on the margins, but absent from the board entirely.

Consciousness was a verboten scientific subject for most of the twentieth century. It took two Nobel Prizes to break it into acceptable discourse, though neither prize had anything to do with the brain or its workings. The first prize was awarded for discovering the structure of DNA, and one of the recipients in 1962 was Francis Crick. The second prize was for discovering the structure of antibodies, and one of the recipients, following ten years behind in 1972, was Gerald Edelman. As their fields moved on, Francis Crick and Gerald Edelman became interested in other disciplines. The terrains they had been the first to truly traverse, molecular biology and immune function, had been drawn out, the boundaries established, and they, explorers, found themselves surrounded by mere cartographers. They yearned for terra incognita, for the empty spot on the map of science, and the greatest gap in the map was consciousness.

So both the American Gerald and the Englishman Francis sought new beginnings in California. And both chose San Diego; Gerald Edelman moving the organization he founded, the Neurosciences Institute, and Francis Crick at the Salk Institute, which he had consulted in founding. Despite their physical proximity and shared interests, the two were closer to rivals than collaborators. Notorious for their larger-than-life personalities, the men clashed personally and professionally over decades. Yet both faced much larger challenges from the scientific community in their struggle to get the study of consciousness accepted as being within the realm of science.

All the way to the 1990s, research into consciousness was not viewed as "proper" science. Yet in their beginnings, both neuroscience and psychology claimed consciousness directly as the phenomenon they were interested in. William James's 1890 field-inaugurating *The Principles of Psychology* coined the term "stream of consciousness." But by the 1920s these approaches had become heavily criticized for their reliance on introspection. Destined to become the most prominent critic was Harvard student B. F. Skinner, although as a young man, when Skinner left academia to live with his parents, he had been determined to write the next Great American Novel.

(How ironic then that, of all cultural products, novels have that special relationship to consciousness. Eventually, maybe inevitably, Skinner became disillusioned with his literary skills, and decided to reject consciousness from science itself. Perhaps the history of consciousness research would have been quite different had Skinner taken to the new extrinsic medium of film.)

Returning to research in psychology Skinner soon became the most significant proponent of a radical form of behaviorism, arguing that organisms were merely input/output black boxes and any talk of the intrinsic perspective was unscientific, barely better than paganism or astrology. If there is a villain for consciousness research, it is Skinner: failed novelist, rejecter of the intrinsic perspective. Due to the popularity of his approach consciousness became a pseudoscientific word and psychology was stripped of the idea of a "stream of consciousness," stripped of everything intrinsic, for almost a century. In order to survive as a science, psychology only kept the reduced elements of consciousness—attention, memory, perception, and action—while throwing out the domain in which they exist, the very thing that gives them form, sets them in relation, and separates one from the other.

In response to a skepticism that had governed entire generations of scientists, both Crick and Edelman fundamentally made the same argument: that consciousness is a natural phenomenon produced by the brain and therefore falls within the purview of science. However, their strategies differed just as the men themselves differed in temperament. Wary of a philosophical morass and a repetition of the rise of behaviorism earlier in psychology's history, Francis Crick argued for the pursuit of what he called the "neural correlates of consciousness," making clear the business of the field was not metaphysical debates but rather correlating changes in consciousness to changes in brain states. As Crick advocated in his book *The Astonishing Hypothesis*,[1] neuroscientists should proceed humbly and empirically, as if compiling a great ledger that maps brain states to conscious states. Whether there was some extra work to be done after the construction of this ledger was a matter for another day; first, the task at hand. Correspondingly he spearheaded a mainly empirical approach, with almost toyishly simplistic hypotheses for what the neural correlates of consciousness might be. He first proposed that neural correlates of consciousness involved the gamma band of brain-wave activity in the visual cortex (neurons oscillating at a particular frequency).[2] The subtext was that what mattered to Crick was not so much that the first scientific hypotheses about consciousness explained what people wanted to know, but that they were testable and set a clear foundation for empiricists to build on.

Gerald Edelman took a different approach. He had always been a fan of grand theories, particularly quantitative or formal ones, that explained some natural phenomenon. This kind of thinking, after all, had won him the Nobel. The immune system faced a seemingly impossible paradox: Given an infinite variety of viruses and invaders always mutating and changing, how could the immune system possibly be preprogrammed with them all? Edelman's theory helped reveal the answer: the immune system has huge reservoirs of randomly generated antibodies floating around in very small numbers, far too small to make a difference or stop an invasion. But the moment one antibody binds to an invading force, the immune system goes through a process of selection that stops the development of other antibodies and massively boosts the production of the antibody that can fight the invasion. Selection from a pool of variability. The forces of Darwinism work in our very bloodstreams. It's an elegant theoretical answer. Edelman hoped to find an equally elegant answer for consciousness itself.

First, he looked in a similar place, positing that the brain worked via a form of selection. He pointed to the immense individual diversity of brain connectivity at birth. A baby's brain is far more connected than an adult brain. Out of the enormous multitude of patterns formed at birth, a subset is selected by being reinforced through learning, while the rest are pruned away to emphasize the lucky selected patterns. These surviving patterns become concepts, behaviors, and other building blocks of the mind.[3]

As what he eventually called "neural Darwinism" developed, Edelman began to link the theory to consciousness. In his book

The Remembered Present (the title comes from William James's description of consciousness) Edelman argued that the set of concepts in the brain, forged by neural Darwinism, were engaged in a kind of back-and-forth interaction; and as these dynamics were categorized into memory, this formed a kind of "remembered present," and it was what we called "consciousness."[4] The theory was critiqued by Francis Crick, who gave it the derogatory name "neural Edelmanism,"[5] furthering the animus between the two men. Later, Edelman's work shifted to examining the complexity of neural activity, helping develop mathematics to directly assess the level of consciousness, again taking a theory-first approach, as he focused more on the dynamics and less on anatomical organization.

Within the space opened up by the Nobel winners, a science of consciousness began to take root. In the mid 1990s Edelman and Crick, along with other scientific heavy-hitters interested in consciousness, like physicist Roger Penrose, had sparked enough interest for there to be an annual conference called *Toward a Science of Consciousness*. Penrose was sure that an answer to the mystery of consciousness must be located in physics, proposing that the brain somehow mysteriously made use of quantum superposition to solve what would normally be impossible problems. Yet from the beginning the inherent tensions of the field were present in its struggle to mark the boundaries between science and pseudoscience. Roger Penrose's group, in its focus on quantum physics, would go on to dominate the conference, until after two decades its worst critics accuse the gathering of now being a summer retreat where Deepak Chopra goes to sign autographs.

To this day most funded and sanctioned consciousness research is populated by the intellectual descendants of Crick and Edelman, and those who worked directly under the two continue to define the field. Because of their influence, consciousness research has been mostly split into the empiricists (the house of

Crick), who focus on brain imaging and finding the neural correlates of consciousness, and the theorists (the house of Edelman), who make quantitative and formal proposals to measure the content and level of consciousness. Members of both these houses are tenured professors at major research universities and receive funding to work on consciousness from the National Science Foundation, the National Institutes of Health, and even the Department of Defense. Neither Crick nor Edelman had achieved the second world-shattering scientific breakthrough he originally sought; instead, the breakthrough was placing consciousness once again under the purview of science after almost a century of absence.

With the field of consciousness research inaugurated, Francis Crick died in 2004 from colon cancer. Again running ten years behind his Nobel-winning rival, Gerald Edelman died from prostate cancer in 2014, following Francis to that ultimate empty spot on the map.

The House of Crick

The house of Crick asks: What is that ledger of correlations between brain states and particular states of consciousness?

First, there are two dimensions to consciousness: the level of consciousness (awake, dreaming, absent, etc.) and the specific contents of consciousness (visuals, sounds, memories, etc.). One of the central concerns of the house of Crick, and also one of its main difficulties, has been to isolate the content of consciousness as an experimental variable (the level of consciousness is much easier). In the past three decades a subtle game of isolation and controls has played out in the literature, attempting to pinpoint the unique difference between when stimuli are processed consciously versus unconsciously in the brain.

For example, let's say you show a perceptually bistable image, like the popular optical illusion the Necker cube, which "flips" when you look at it long enough. You then tell a participant to left click for when they see the higher plane of the cube "in front" and right click when they see it "in back." Just this act of clicking right when seeing one image and left when seeing another means that differences in the parietal and motor cortexes of the brain are thrown into the data, since clicking left requires different motor area activity in your brain than clicking right. And worse yet, participants have to make a conscious cognitive choice to click a certain way. So are you measuring this choice or the perception? In other words, the report of consciousness itself interferes with isolating consciousness as a variable. This means there is an observer effect: to get data on consciousness we need report, and yet report interferes with isolating consciousness.

Enter "no-report paradigms." In these cases the experimenter relies on something automatic that is first shown to be tightly coupled with verbal report, and then eschews report altogether once a coupling is established. For instance, a classic illusion that switches between one perspective and another can be first correlated via report to subtly different pupil reactions in the participants when viewing each of the perspectives. Then the subject doesn't even need to report when perspective on the illusion changes in order for investigators to know what's happening in the subject's brain during such changes in perspective. Ironically, one main takeaway of this game of controls is that the neural correlates of consciousness are relatively local. The visual cortex seems to process visual conscious experience, the auditory cortex processes sounds, and so on. The more you control for, like report, the smaller the footprint of consciousness gets.[6]

As an undergraduate I had the run of an EEG lab to do experiments in, and I wanted to replicate a study that had used almost every control possible to isolate consciousness. The claim of the study was that only the prefrontal area of the brain showed a difference between unconscious and conscious stimuli once you controlled for performance on the task itself.[7] But that study had imaged the participant's brains using fMRI, which has very poor temporal resolution, so I wanted to use EEG, which has orders of magnitude better resolution in time. Unlike the original study, due to our different methodology, we saw very clear differences in the posterior part of the cortex (the visual cortex). First the subject saw a dot, and then the flash of one stimulus (a square) and then another stimulus (a diamond) that masked it (so sometimes the subjects saw and sometimes they didn't see the original stimulus), and the activity in the visual cortex reflected this perfectly. Yet the paper was being drawn up for publication; I also realized that no matter what I did there would still be a variable left uncontrolled: in the case of a successful mask, the subject saw a dot and then the mask, and in the case of an unsuccessful mask, the subject saw first a dot, then the stimuli (square or diamond), then the mask. In the language of William James, the former case would be a stream of consciousness that contained two successive perceptions, one after the other, and the latter case contained three. I wasn't comparing one conscious perception to another unconscious one, I was comparing seeing two things in a row to seeing three things in a row! So my coauthors and I never published the results, despite the fact that hundreds of published papers have used similar methods to make claims about consciousness. A question has haunted me ever since: If consciousness is a stream, then it is the medium in which other psychological processes take place, and is it ever possible to isolate a medium?

It is because of problems like this that the search for the neural correlates of consciousness will likely not be driven by a simple experimental revelation, by which I mean some discovery that leads naturally to a particular theory. One can see why Francis Crick would have expected this—after all, he and James Watson ended their famous paper elucidating the helical structure of DNA with the oft-quoted understated acknowledgment that "it has not escaped our notice that the specific pairing we have postulated immediately suggests a possible copying mechanism for the genetic material." That is, the empirical evidence of the structure of DNA was immediately suggestive of the broad theory of DNA function—one followed logically from the other.[8]

But this is unlikely to be the case in the brain. We know some extremely broad facts with moderate certainty, like that the difference between consciousness and unconsciousness involves how complex neural activity is.[9,10] But this sort of broad association does not give us any specific insight into consciousness. And if the goal of the house of Crick is simply to create as detailed a map as possible, without looking for an experimental revelation, this runs into another problem: the neural correlates of consciousness then have their own difficulties *plus* all of the problems of regular, pre-paradigmatic neuroscience (which in turn, has all the problems of psychology). Across 412 studies examining four different theories of consciousness, the methodologies were widely diverse, and the research almost always found only confirmation of whatever theory the authors were advocating for.[11] It is for this reason that the search for the neural correlates of consciousness has made little progress—it is as impossible to use pre-paradigmatic science to trigger a paradigm shift as it is to pull yourself up by your own bootstraps.

Not only that, there is a sense in which empirical investigations render the whole thing more difficult, for any theorizing must first decide what experimental evidence to take seriously, and many of its findings contradict one another, meaning that empirical results, rather than pointing the way to an obvious hypothesis about consciousness, instead create a mass of choose-your-own-evidence and empirical contradictions.

And just in general, can a purely empirical science really exist by itself, shorn of theory? That is, via correlation alone? So far the fields that have lacked clear overarching theories, like sociology and psychology, have been the ones that have failed to transform the lead of academia into the gold of science.

Which leaves only the house of Edelman.

The House of Edelman

Edelman's Nobel Prize was based on a different direction of action than Francis Crick's: rather than a mechanism suggesting a theory, in Edelman's work a theory suggested a mechanism, which Edelman then validated.[12] According to his wife, his own PhD advisor didn't believe in Edelman's theories, and therefore did not end up sharing the prize with him.[13] We see, in other words, the antithesis of the Crick style of research, one where the right theory is necessary to unlock the right empirical results, rather than the other way around. It is unsurprising that the house of Edelman continues on in the work of Edelman's students, who include big names (at least within neuroscience) like Giulio Tononi, Olaf Sporns, and Karl Friston (the most cited neuroscientist), among many others.

Due to its theory-first roots the house of Edelman asks: How can one craft a theory of consciousness? Note that crafting scientific theories is an art: theories must have certain characteristics, including parsimony, explanatory and predictive power, elegance, and ultimately they should be suggestive of how to prove or disprove the theory experimentally. Theories often have multitudes of moving parts, and constructing one can be more akin to solving a sliding puzzle rather than answering a single riddle.

Operating at a more abstract level, the house of Edelman mirrors the relationship theoretical physics has to applied physics. Less like the search for the neural correlates of consciousness, with its focus on the tools of standard neuroscience like brain imaging and quirky neurological cases, Edelman's house is a place mostly of abstractions, thought experiments, and mathematics. Yet it is the only place that has a chance of moving neuroscience to its necessary post-paradigmatic state.

The dream of the field can be summed up by the idea of a "conscious-o-meter." Imagine a tool that would tell you, for any given system (like an awake human brain or an artificial neural network), if it is actually conscious, and what it is conscious of. More generally, we can describe this abstractly as looking for a mathematical function that maps physical states onto mental ones. My colleagues and I have outlined what such an ideal theory of consciousness would look like and attempted to do so as rigorously as possible,[14] which is useful as it allows us to talk about theories of consciousness without committing to a particular one. So what precisely does a theory of consciousness look like?

A theory of consciousness, at least the kind nigh universal in the current field, starts with some p, which is some configuration, dynamic, or state, of a physical system P. In most relevant cases, this physical system would be the brain, but when thinking of things formally we may as well make it as general as possible to cover any edge cases and say it can be any given candidate physical system. In this formalism, p might be a brain state (or dynamic or configuration; language is loose here) and P is the brain itself. This is written $p \in P$, where p is a member of the set of possible configurations or neural dynamics or brain states, whereas P represents the total set of possible ones. Furthermore, there is some set of relevant observables (data, variables, things we're paying attention to) in the theory, which might be neuroimaging data. Let us call this o, wherein o is some member of the set of possible data (this is written $o \in O$). There is some mapping from $P \rightarrow O$, an arrow we can call "*obs*" as short for "observations." This reflects the standard method of science finding observables of interest from some physical system and recording them or analyzing them.

The target of a theory of consciousness is the set of various different conscious experiences a system could have (each theory carries with it its own definition of this space). Within the set E there is some particular experience that the system is having, e, wherein, again, $e \in E$ (your current e is reading this book, perhaps it also includes drinking coffee or sitting in "the comfy chair"). A theory of consciousness is some mapping that predicts what experience the system is having (out of the space of possible experiences); let us call this "*pred*" for short.

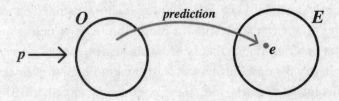

Broadly, we can think of theories of consciousness as a chosen mapping between a physical system (or observations about it, O) and the experiential (E). P and O exist in the extrinsic perspective of the world, that of science and mechanism, that of states and dynamics. E exists in the intrinsic perspective of the world, the set of possible experiences. What a theory of consciousness does is map extrinsic data onto the space of possible experiences E. It is a way to traverse, or at least lawfully relate, the extrinsic perspective to the intrinsic. And a theory of consciousness is "simply" knowing which function to use!

Consider one of the earliest, but still very influential, theories of consciousness: that consciousness is a kind of global workspace for the brain. First introduced by the cognitive scientist Bernard Baars in the late 1980s,[15] and grounded in neural dynamics by scientists like Stanislas Dehaene in the late 1990s,[16] the theory rests on a metaphor: consciousness is the bulletin board of the brain. Local parts of the brain start "shouting" at the rest of the brain, and the processing or information that is shouted out loudly enough to be global is conscious, whereas signals that are kept local are not.

But is global workspace even a full theory of consciousness? When exactly are things in the workspace, and when are things out? What are the minimum requirements to have a global workspace? Can the answers to any of these questions be stated mathematically? All of these problems can be summed up as: given some arbitrary p, how do we observe it (O) and then how can we map those observations into the space E, according to global workspace theory? And if there isn't a relatively clear answer to this, isn't global workspace more like a metaphor than a theory?

Just as it would be unacceptable to physicists to have a theory of gravity that was merely metaphorical but not mathematical, those in the house of Edelman turn toward the formal and the quantitative (and indeed, I should mention that later versions of global workspace theory have become even more formal and quantitative[17]).

The star of the house of Edelman is a full and complete theory of consciousness. Possibly the only theory that even meets this requirement to this day was proposed by neuroscientist Giulio Tononi, who was a postdoc under Gerald Edelman at the Neurosciences Institute (and I, in turn, was a graduate student under Tononi, making Edelman effectively my "scientific grandfather"). Tononi spent ten years with Edelman, and right after Tononi left the institute and working with Edelman, he introduced Integrated Information Theory (IIT).[18] The neuroscientist Christof Koch, president of the Allen Institute for Brain Science and a close collaborator of Francis Crick, famously called IIT "the only really promising fundamental theory of consciousness."[19]

Why does IIT deserve such laurels? First, its theory-first approach has led to a number of quite intriguing and interesting empirical results by identifying consciousness with how information is integrated in the brain, and the complexity of the brain's internal relationships. While the true calculation of integrated information in the brain is impossible, there are some clever heuristics that have been developed: specifically, the perturbational complexity index, which is based on how complex cortical activity is following a perturbation of a bunch of neurons using an electromagnetic field. In other words, you ring the bell of the cortex and see how pretty a sound it makes. Which means even for a patient with locked-in syndrome we'd be able to get a sense of how conscious they actually are (note the moral importance of this question, as knowing their state of consciousness might mean just doing something simple like making sure the TV or radio is on for the poor soul stuck in the Tartarus of their body). And indeed, awake patients have the highest perturbational complexity, but locked-in patients have a higher perturbational complexity than minimally conscious patients, who in turn have a higher perturbational complexity than the fully unconscious.[20] The same is true of patients with subcortical atrophy, wherein the degree of atrophy is tracked by the perturbational complexity of the cortex.[21] And while highly expensive and technical means of cortical perturbation, like using transcranial magnetic stimulation (a focused magnetic field), is not available for most clinics, "natural" perturbation is possible via administration of propofol (an anesthetic)—and, in one study, this indeed predicted which coma patients would recover consciousness.[22] The perturbational complexity index, since it is a heuristic, is not the direct calculation of integrated information, but it does tell us that consciousness might be related to cortical complexity in general, and the measure, so useful for evaluating comas and minimal consciousness in patients, is a direct intellectual consequence of the intuitions behind the theory itself. This is a sign of a healthy theory: that it leads us to empirical ideas we wouldn't have had without it.

But IIT is still far from experimentally confirmed; there are many debates surrounding it, its current formulations are controversial, and even its adherents openly admit that it's a work in progress. What IIT asks is this: Given what we know from phenomenology, the structure of our own conscious experience, how is this mapping between the experiential and the physical constrained?

The reason for its popularity is that IIT is the first real model of what a scientific theory of consciousness might actually look like. Even to this day, it is perhaps the only theory of consciousness that satisfies the definition of a theory of consciousness given above: it allows you to say, for any given p, what e will be.

And that is why I, at the age of twenty-one, back in 2010, went to go help develop it. At the time it was even less known than now, just a handful of papers really. The IIT group under Tononi was the only place to study a theory of consciousness seriously in the entirety of the United States, which tells you something about the scale of this field (minuscule).

To become a scientist you must interview at various graduate schools, in a way that anyone familiar with the normal admissions hoops of academia will recognize. An important part is the interview with the professor whose lab you want to work in—and when I applied to the University of Wisconsin, it was solely to work with Giulio. I had no interest in anything the other professors were doing. But when I went to go for my interview weekend, there was a "snowpocalypse" on the East Coast and all the flights were grounded. I missed it. Fortunately, the small number of us who were grounded were instead invited to a different week. I attended it and met with Giulio.

In our conversation, a little-known philosophical principle came up from a now-little-known philosopher at the turn of the twentieth century, Samuel Alexander. The principle was Alexander's Dictum ("only that which causes exists"). Causation plays an important role in IIT, and Giulio had been reading about it. I had also heard of Alexander's Dictum before, having stumbled across it on one of my many nights in the library as an undergraduate, browsing late through the stacks and reading alone at the carrels. Giulio had never run into anyone outside himself who knew of it. He essentially hired me on the spot, and asked the program to officially admit me, which they promptly did.

Later, I found out that on the original weekend of the snowstorm, Giulio wasn't in town. See, graduate schools don't normally assume that students must talk to someone in particular—that level of specificity is abnormal. So they didn't even check to see if Giulio was around. I would have missed the only person I actually wanted to interview with. Even if the program had accepted me, which is doubtful, I myself wouldn't have accepted, as I wouldn't have been able to even talk to the person under whom I wanted to spend the next five years working, let alone get him to vouch for me.

So I got into consciousness research because a butterfly flapped its wings. Which caused some microscopic turbulence, which chaotically spun round the globe, until finally thick flakes fell over Boston, Massachusetts.

Only that which causes exists.

Phenomenological Theories of Consciousness

Far away in New Zealand, and more than a hundred years ago, in a foreshadowing nearly lost to history, an eccentric named Benjamin Betts began to dream of the first mathematical theory of consciousness. He imagined a system very much like a contemporary theory of consciousness from the house of Edelman, a mapping between phenomenology and some mathematical system, which for Betts took the form of a dense and arcane language of geometric shapes. Betts had an odd background, not that of a scientist; he had been educated in England and later spent time in India, studying Far Eastern philosophy.

In 1897, his musings were collected into an eccentric monograph of drawings, equations, and philosophical tangents and published as the book *Geometric Psychology*. His sister just happened to be a friend of the widow of George Boole, one of the great mathematicians of the time. Boole's widow popularized Betts's work within her own scientific circles and corresponded with him, and eventually his work was collected and edited by Louisa S. Cook into a volume that survived all the way to the present day.[1] According to the collection put together by Louisa: "Mr. Betts felt that consciousness is the only fact that we can study directly, since all other objects of knowledge must be perceived through consciousness."[2]

This is, of course, correct. However, his system quickly spun into fancy, leading to wild statements like "The symbolic representation of animal sense-consciousness is in two dimensions, and in form resembles a leaf whose apex is about equal to a right angle."[3] Yet, despite the specificity, of equating animal sense-consciousness with a leaf whose apex is about equal to a right angle, Mr. Betts never justifies such mathematical representations—for his proto-science of consciousness he offers no justifications, simply that it was what intuitively made sense according to his own quirks. His theory, difficult to decipher, appears to be a panpsychist one, with consciousness as Being itself. In the cases where there was justification of the math used, it was too metaphysical, drowned in Betts's personal philosophy and musings on the nature of Being and spiritual activity.

Which all makes *Geometric Psychology* fascinating but, I should be clear, not a traditionally scientific text. It contains a lot of woo-woo, like "Love is the Substance of all things"[4] (unless, of course, love turns out to be the substance of all things, vibrating under the atomic). His system is ultimately unworkable—a loose collection of metaphysics that only amount to something suggestive. But despite all those caveats, as we will see, in some interesting ways it prefaced contemporary theories of consciousness, and so contains an early foreshadowing of the approach favored by the house of Edelman. It is the first attempt that I know of to represent qualia mathematically, and many of its metaphysical points are later recapitulated in Giulio Tononi's IIT. Here are some of Betts's sketched representations of conscious experiences as geometrical objects:

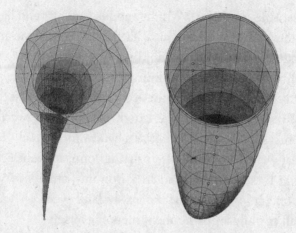

This sort of proto-scientific exploration of the intrinsic perspective via the extrinsic one was "in the air" around the turn of the century. Like the work of Wilhelm Wundt, one of the most influential early psychologists. Some historians of science credit him with single-handedly founding the field of psychology itself (he's often dubbed "the father of psychology"). Little known now but a giant of his time, Wundt overlapped Betts, and by 1875 he was a professor in Leipzig in Germany and on the way to becoming famous for transforming psychology into a science. He was a humorless, truculent, and stubborn man, with the work ethic of a draft horse. In a damning accusation, William James described him in a letter to a friend this way: "He isn't a genius, he's a professor."[5]

Wundt's original conception of psychology involved a great deal of careful and logical introspection. He argued that the job of the psychologist was to take the intrinsic perspective and both examine their own consciousness and teach others to do the same, reporting specifically on the "elements" of experience, which for Wundt meant the simplest possible sensations, the buildings blocks of consciousness. He wanted to create an intrinsic mirror of the periodic table, examining all our minimally distinguishable sensations. Wundt and his student (and later scientific standard-bearer) Edward Titchener used experimental psychology to compile thousands of simple sensations,[6] eventually reaching more than forty-four thousand of them across the different sensory modalities (these included things like the duration of conscious events, the extent of objects of perception, hues, colors, sounds, etc.). This was a heady intellectual time and these ideas were in the air. In Leipzig the young student Edmond Husserl, at the time interested in astronomy, attended Wundt's lectures on philosophy—and famously later went on to try to ground all of philosophy in phenomenology itself.

Skinner would later argue that Wundt and his disciples were promoting unfalsifiable ideas, and under the assault of behaviorism the notion of a scientific mapping of phenomenology would fade into history.

Until it would be revived in the house of Edelman. For Integrated Information Theory is itself based on introspection, much as Wundt's work was. The result of this introspection is a set of axioms: indubitable statements about conscious experience. The approach is modeled on Euclid's *Elements*, the most famous book of mathematics, in which two thousand years ago geometry was formalized based on a set of five axioms, statements like "It is possible to draw a straight line from one point to any other point."

IIT aspires for the same clarity, although there is room for debate if it's been reached (more on this later, for now let's take the theory as is). For a phenomenological axiom to be an axiom, it must possess certain irreducible characteristics, and therefore it must not be derivable from some other axiom, it must be universal (it must be true of all experiences), and it must be as simple as possible. In the current version of IIT, there are five axioms, just as in Euclid. See how much you agree with them, based on your own introspection:

1. Consciousness exists. (Note that this, IIT's first axiom, is very close to the first law of Benjamin Betts's *Geometric Psychology*, which is "Being shall exist."[7])

2. Consciousness is informative in that you are always having a specific experience, and that experience is different from any other experience—you might be seeing a particular frame in a movie, which is different from all other frames from movies.

3. Consciousness is composed, in that it has a certain structure and variety—you have a left side of a visual field, a right side, and so on, and the elements of your stream of consciousness often relate to one another. (The second law of Benjamin Betts is "Being exists in variety."[8])

4. Each experience is integrated. Your consciousness is a gestalt irreducible to its component elements (if you are looking at a red cube, it is not like looking at merely redness plus a colorless cube).

5. Consciousness is definite. This means that your consciousness is one particular way, not another—it also means that your consciousness has a certain temporal grain. It flows at a particular speed, and contains particular elements, and whether you are aware of certain things is an answerable question.

While I was getting my PhD, the small team I worked with advancing IIT would discuss these axioms endlessly, hoping to distill them down into something as crystalline as Euclid's *Elements*. And in many cases they do seem to be intuitive, at least to most—your consciousness definitely exists, and is composed with some sort of internal structure, and is obviously informative, using at least a broad definition of "informative" (your conscious experience is one among the astronomically many you could have, and also contains information about the world and yourself). The relationship between integration and composition may involve some confusing and not obvious overlap, and the last axiom, definitiveness, is the least innately obvious to most. For now, let us assume it is plausible that these axioms are true of all conscious experiences—at least, while we overview the theory.

In 2004 Giulio Tononi published the first paper on what would become Integrated Information Theory.[9] While the theory did not originally start as a mapping of phenomenology to physical states, nor was it originally based in axioms, it was one of the first mathematical measures of consciousness, having been born of work that Tononi did with Edelman himself on measuring the neural complexity of the cortex as a potential barometer for consciousness.[10]

When I arrived, although they had not been fully set yet, Tononi had already advanced the idea of grounding the theory in phenomenological axioms.[11] Only two of them had been proposed, those of information and integration (thus "Integrated Information Theory"). One reason I found IIT so attractive was that I had done my undergraduate thesis on compiling the properties of consciousness, following without knowing it a similar approach, writing in my undergraduate thesis that

> one must examine the properties, the structure of experience, and as one lists the properties the possible explanations for consciousness will be reduced, as each property is naturally limiting. In large combinations, these properties fix any theory of consciousness within a limited range.[12]

After I joined the lab, the list of axioms of IIT eventually expanded to the full five, and the math of the theory itself grew more complex.

Where does this math come from, given that the axioms don't

immediately suggest any equations? Overall, the goal of IIT is to look at physical systems to see whether they satisfy these axioms (like whether or not the physical system is integrated) and therefore identify if they are conscious and to what degree. So unlike in Euclid's axioms, which are already directly stateable in the language of geometry, IIT has the further complexity that there needs to be a mathematical mirror of the phenomenological axioms, like a translation from one language into another. Since IIT's axioms are stated in the language of the intrinsic, they must be translated into the language of the extrinsic.

To do this, IIT grounds itself in an information-theory-based analysis of causal models. The why behind this choice is a good question, and is actually based on the first axiom: that consciousness exists. And what is it to exist? Well, to exist is to cause. Remember Alexander's Dictum: only that which causes exists. From this point of view, only causal models are "real"—while we can give non-causal accounts of the universe, such as how two things are correlated or associated, these are merely epiphenomenal shadows of the actual causal account of events.

This may seem an extremely metaphysical and philosophical reason to ground the theory in causal models, and there is no denying that it is. But we should be open about the fact that any good theory of consciousness is likely going to have metaphysical implications and assumptions.

There is an additional benefit of this choice: by basing its calculations on something as abstract as causal models, even if IIT turns out to be an incorrect theory of consciousness (a) why it might be incorrect is quite interesting and tells us about general pitfalls a theory of consciousness will run into, and (b) luckily, IIT tells us other facts about how to partition and understand complex systems above and beyond questions about consciousness.

What precisely does IIT mean by a causal model? Causal

models are just descriptions of how parts of systems relate. There are all sorts of technical differentiations with regards to causal models, but they are basically always "dots" (things) connected by "arrows" (relationships) of some kind. The study of causal models is an abstract mathematical field, but that abstractness has a purpose, which is to be general enough to represent many things. A model of a microprocessor is a causal model, where the elements are transistors and the way they are wired together are the arrows of influence. A causal model of the brain would look like the set of neurons and their weighted connections (how much each neuron can influence other neurons). To see how these are related, we can further assume that the neurons are "all or nothing" in their activity, that is, at any time-point t they are either firing an action potential or not, which, in simplified form, looks an awful lot like a system of simple binary logic gates (also known as a Boolean network—after George Boole). Therefore, just to be as clear as possible, we can specify the mathematical interpretation of the phenomenological axioms in the "mini-brain" that a system of logic gates represents, along with their causal relationships. And we're going to represent them "dynamically," that is, so we can clearly see the inputs to different parts of the system and their outputs at the same time. Of course, the assumption that we are dealing with such simple systems might have consequences later on, but at this time, let's detail the sort of system for which we want to calculate the integrated information for to look like this:

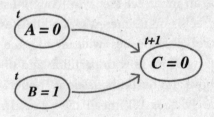

This sort of causal model is actually quite easy to understand. We have three "elements" or "nodes" or (if we're using the brain analogy) "neurons." They are labeled [ABC] and each element is in some state. Assuming that these elements are binary, the state could be either 0 or 1 (above, A = 0, B = 1). Both elements A and B input their current states to C at time t. C has a truth table by which it makes some mechanistic decision about its own state (note that this happens at t+1, the time step immediately following t). C could be following any number of truth tables, like one of those below. The way to read a truth table of elements in a causal model is that there are inputs (on the left), and then outputs (on the right).

A, B	AND
00	0
01	0
10	0
11	1

A, B	XOR
00	0
01	1
10	1
11	0

A and B might have truth tables, too, and have their own inputs—this could only be a partial view of a greater system. But we can leave aside those details for now and focus on C. Let's assume that C is following an AND truth table (the one on the left above). If A inputs 0, and B inputs 1, which is {01}, then C, since it is following the AND truth table, responds by assuming the state 0. That IIT is analyzing such causal models of mechanisms in particular states is the physical interpretation of the phenomenological axiom that consciousness exists.

IIT's second axiom, that consciousness is informative, can also be given a physical translation: the mechanisms of a causal model are informative, since their states contain information about the past and the future. E.g., if we know that A = 0 and B = 1 at time t (the present), then we know that C is going to be 0 at t+1 (the immediate future). This goes backward, too: if we know only the state of C, we can come to conclusions about A and B's previous states. Imagine that we knew only the information that C = 1. If so, we could look at the truth table of C and conclude that A = 1 and B = 1 (the circled row of the truth table below) must have been true the moment before.

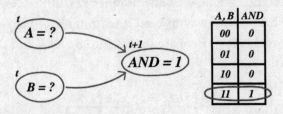

This is the information part of Integrated Information Theory. IIT quantifies the amount of information that the states of elements have about one another: here, the AND has two bits of information, since it can tell us the answers to two yes-or-no questions, which are what the states of A and B had to have been. While there are further details about precisely how this information is measured, as you can do it in a lot of ways[13] (indeed, it changes version to version of IIT), this general concept is the heart of the information part of the theory.

The third axiom, that consciousness is composed, can be translated as meaning that information itself has a certain composition. In the causal model, looking at A's current state gives information about the future state of the AND, but so does looking at B, and so does looking at the full set of both inputs {AB}. The question that IIT asks is whether these different sets of information are reducible to other sets of information. How can we assess this? Simply by examining the information derived from different states of elements and judging whether that information overlaps and is redundant or not. This is best seen through example. Let's say we want to assess whether the information that {AB} contains about the future state of the AND is reducible or irreducible. To judge this, we can see if there is overlap with the information its parts contain, such as A. For an AND, if the inputs {AB} are {01} at t, then we know that AND = 0 at t+1 (the circled part of truth table).

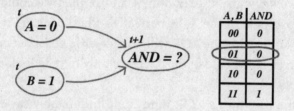

Compare this to if we only knew the state of A, and the state of B was unknown. Knowing only the state of A, we actually can predict again that AND = 0, which is the exact same information.

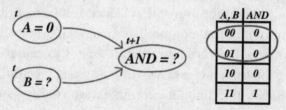

Thus, we can conclude that the information {AB} contains about the future state of the AND is totally reducible to just the information that A contains alone.

What would be a case of irreducibility instead? What if this weren't an AND, but instead, an XOR? This would mean it has a different truth table. If it were an XOR (a truth table called an "exclusive or") wherein the mechanism outputs 0 if and only if it receives two of the same input (whether they be 0 or 1). In this case, knowing only that A = 0 does not tell us anything about the state of the XOR at t+1 (and the same goes for B).

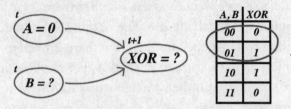

Yet, if we knew the state of both A and B, then we could predict the future state of the XOR perfectly. What it means is that we cannot know what the XOR will be unless we observe *both* A and B, whereas in the case of the AND we only have to look at one of its inputs.

In the case of the XOR, one could indubitably say that "the whole is greater than the sum of its parts."[14] Yet prominent scientists will regularly say the exact opposite. Here is Sabine Hossenfelder in her recent book *Existential Physics: A Scientist's Guide to Life's Biggest Questions*: "As a particle physicist by training, I have to inform you that the available evidence tells us that the whole is the sum of its parts, not more and not less."[15] Perhaps this is somehow true, but certainly when it comes to causal models, whenever the truth tables that govern a system are non-additive and nonlinear, by definition information won't always be reducible to the sum of its parts.

This is the heart of the concept of irreducibility in IIT, although, again, there are some further complexities and details in its implementation. This process must be done exhaustively (so that every possible set of elements and their information is calculated) to compile a huge list of irreducible information (given the symbol "small phi" or φ), which might look like A's $\varphi = 0.5$ bits, B's $\varphi = 0$ bits, {AB}'s $\varphi = 1$ bit and so on.

The fourth axiom, that consciousness is integrated, has an obvious physical translation—whether causal models are themselves integrated or not. This can be ascertained by partitioning the model and asking how much irreducible information is lost (the type of information specified by the second and third axioms). Notably, this differs from the previous axiom of compositionality, in that it is asking to what degree the system as a whole is integrated, rather than how irreducible the information gleaned from the elements themselves is. In IIT this is given the symbol "big phi," or Φ. The way this is assessed is to see how much damage to the information occurs given a partition, like a literal cut between elements of a system. This is conceptually like asking: how much information in the system is lost if we severed A off from the rest of the system? What about B? What about both? And so on.

Assessing the amount of "damage" to the irreducible information in the system that severing parts of it does (sometimes called "partitioning" or "cutting") entails finding the severing that does the least damage possible. Why do we want to find the "cut" that does the least damage? Because we can use it to draw boundaries between systems and their presumed consciousnesses. This is a thought experiment originally proposed by William James:

Take a sentence of a dozen words, and take twelve men and tell each man one word. Then stand the men in a row or jam them in a bunch, and let each think of his word as intently as he will; nowhere will there be a consciousness of the whole sentence.[16]

IIT would be uniquely able to distinguish this case, for a partition between the brains of the twelve men would reveal that no information is lost—conceptualizing the men as separate entities leads to no loss in total information. Therefore, it wasn't an integrated system to begin with. Again, this is the high-level conceptual presentation, as there are further details. IIT must specify some metric by which we can judge how much difference a particular cut makes to a system. In IIT the metric when I was working on it was the "earth mover's distance,"[17] which measures how much "work" it would take to transform one probability distribution into another (basically by just asking: How minimally does distribution 1 need to shift to be transformed into distribution 2?), and then assesses how much "damage" is being done to the information after the cut.

Additionally, all these things must be calculated for *every* subset of the system, e.g., both {ABC} and also {AB} and {BC} and {AC}. The problem is that in larger systems the number of possible subsets grows astronomically large. And every subset will have a different Φ.

The fifth axiom, that consciousness is definite, deals with the issue that almost every part of a system can have some positive Φ. The fact that consciousness is definite is translated physically into the principle of "exclusion," wherein irreducibility and integration are given a winner-take-all aspect. According to the theory, what matters for consciousness is the maximum of integrated information, not just integrated information itself. Thus, rather than finding simply φ or Φ, this axiom tells us to associate consciousness only with φmax or Φmax.

Applying these translations essentially creates an intensive algorithm, the scale of which precludes it from being run on anything manageably. Although it's also worth pointing out that this is merely inconvenient, not directly a knock against the theory—systems don't calculate their own Φ at every time step in the same way that thrown rocks don't need to know calculus.

Axioms and Their Discontents

Several criticisms and questions about IIT present themselves immediately. These break down into two camps. The first camp is whether the axioms of IIT are correct and justifiable. The second camp of criticism concerns whether the axioms, even assuming they are true, are being appropriately translated when applied to physical systems.

The first camp of criticism focuses on the *completeness* of the axioms of IIT. Are there any axioms missing from IIT? This seems like a fundamental question, one which has not been well addressed in the development of IIT. Indeed, there's a commonly noted observation about consciousness that might reasonably be missing from IIT: consciousness always has a perspective. There is a self, an "I," who is experiencing these things. One might reply that there can be states in which there is no "I," perhaps in deep meditation or, say, during an orgasm, but it is arguable that even for the most close-to-enlightenment bodhisattva meditating on nothingness, there is still a self experiencing these things, there is still a point of view.

Yet in IIT as it currently stands there is no center, there is no "vanishing point" inherent to the geometric shapes of the qualia. Betts himself thought that the perspectival nature of consciousness was axiomatic, and that this needed to be reflected mathematically. As he put it:

When we contemplate our consciousness we find there is one element which differs from all the rest. Whereas they are multitudinous, chaotic, changing, it is one, alone, comparatively unchanged. . . . We call it "I," the subject of consciousness. . . . We feel as if our centre were fixed, and so far as its relation to its own activities are concerned it is fixed. The ego is always the centre of the diagram wher- ever that diagram may be located.[18]

Some might even just straight-up deny some of these axioms, particularly the last axiom of the definitiveness of consciousness. One notable example is the popular philosopher Daniel Dennett, who denies the definitiveness of consciousness in some of his work.[19] Furthermore, we cannot rule out the possibility that there are consciousnesses with entirely different axioms, or that are missing some axioms or that have some additional ones that we lack. To give a historical analogy: for millennia Euclid's axioms were taken to be obviously true. There was only one type of geometry: the Euclidean kind. But then in the nineteenth century a number of "non-Euclidean" geometries were identified, in which certain axioms change. These non-Euclidean geometries are often self-consistent and fecund. Indeed, our own world is not really Euclidean (due to the bending of space-time). So might there not be "non-Euclidean" consciousnesses in which several of the axioms are different? This is a bit fantastical, as perhaps people are imagining aliens or advanced AIs, but it's still impossible to rule it out. And we might not have to look far for a non-Euclidean consciousness. Here is Herman Melville's description of whether or not a sperm whale's consciousness is integrated:

How is it, then, with the whale? True, both his eyes, in themselves, must simultaneously act; but is his brain so much more comprehensive, combining, and subtle than man's, that he can at the same moment of time attentively examine two distinct prospects, one on one side of him, and the other in an exactly opposite direction? If he can, then is it as marvellous a thing in him, as if a man were able simultaneously to go through the demonstrations of two distinct problems in Euclid. Nor, strictly investigated, is there any incongruity in this comparison.[20]

Following Melville, it is arguable that our notion of consciousness necessarily being integrated is based on our visual fields being unified; if we were like whales, or rabbits, or other prey animals, we might view consciousness as much more disparately put together. However, the IIT proponent can simply say that there is no reason to posit extra axioms until we are forced to, that they can just be added onto the theory, and that there is no strong evidence at present for "non-Euclidean" accounts of consciousness.

(Incredibly, long after I had picked out that quote from Melville as being demonstrative of this issue of "non-Euclidean" consciousnesses, I had a private conversation with Dan Dennett himself in which he independently brought up that exact same objection and, to my astonishment, referred to the same quote from Melville. I could not figure out a way to tell him I had been using the same analogy without him thinking I was lying; it was simply too fantastical a coincidence.)

Beyond the axioms themselves, there is the question of how the axioms are translated into the extrinsic perspective. Even if we conclude that consciousness is integrated, and agree this is axiomatic, how do we know if a physical system is "integrated" and what, precisely, do we mean by this? IIT certainly has its own answers, but there seem to be a large number of possible choices here—one might argue that neural synchrony across brain regions counts as "integration," and so on. IIT explains why it is that consciousness has borders, but is it the only possible explanation of this? How else might systems get borders beyond their causal integration?

Well, black holes have borders. They have an event horizon from which light cannot escape. And while it might seem a stretch to compare a brain to a black hole, as black holes are event horizons for light, other systems can also have event horizons, for other properties. For instance, sonic black holes are when a fluid flows faster than the speed of sound[21] (imagine here a fish screaming upstream while being forced downriver in a stream flowing faster than the speed of sound). Given that the brain processes information in a manner canonically described as a stream (the "ventral stream" for recognizing what things are, and the "dorsal stream" for recognizing where things are[22]) perhaps there also may be some sort of analogous information "event horizon" past which information cannot flow backwards. This would explain both the definite nature of consciousness (the axiom of exclusion) and its integrated nature (the axiom of integration). My point here is not to offer actually fleshed-out alternatives to the chosen translations of the axioms that can be applied to physical systems, but merely to point out that alternatives are conceivable.

In an ideal case, the axioms of IIT would allow for the deduction of the translations of the axiom directly and independently (as in, each translation would be uniquely fixed and derivable from the axiom alone). Instead, it appears that IIT relies on a broader notion wherein the axioms of IIT may in some cases only *abductively* entail the algorithm (*abduction* meaning "inference to the best explanation," or essentially "best guess"). This has been pointed out by philosopher Tim Bayne, who writes:

> One would need to show not only that the postulates of IIT account for the axioms, but also that they provide a better account of the axioms than competing accounts do. However, the IIT literature makes no attempt to show that IIT does provide the best of the available explanations for the axioms—indeed, other possible explanations of the axioms aren't even considered.[23]

A specific case of this problem is that aspects of the IIT algorithm are not determined by phenomenology at all—certainly, they are not derivable from it alone. For example, according to the theory, joint effects are equivalent to actual psychological concepts, what Wundt would call "elements"—the irreducible building blocks of conscious experience. This is odd, as nothing in the intentionality of consciousness makes it seem as if your concepts are about the past or the future states of your brain, which is the way information is dealt with in IIT—your conscious experiences are always about the state of the world, and while they often seem to be about the state of yourself as well, it's a stretch to say they are a prediction about the immediate future and a retrodiction about the immediate past of your brain state. They simply don't, at least via introspection, appear to refer to that. This highlights that there is some jankiness in the translation of the phenomenological axioms into the mathematics of IIT. To deal with this, it has been proposed that somehow the conceptual structure of the quale "matches" the world, and that this is the basis of intentionality, but it's still unclear what this means, the math behind it, and how intentionality would actually emerge from such matching. Nor does it address what seems like a missed opportunity—to have the informativeness of IIT be related to the intentionality, the aboutness, of conscious experience.

Another example: when considering a subset of a system for the IIT algorithm, one must fix a background condition, that is, an assumption about the state of the rest of the world outside the system while the calculation is performed. To make this clear: calculating the integrated information in the brain requires mathematically "freezing" the rest of the universe in some state, so that all the probabilities and causal relationships can be analyzed. But when doing so you must make the choice: Do you condition over the background elements at time t, or at time $t+1$?

That is, the state of the world one time step before, or at the same time step? In one paper, the so-called IIT 3.0, it was at time t. Later, it was changed to be time $t+1$. What is ensuring the truth of either interpretation of a phenomenological axiom? What, precisely, about phenomenology informs us to translate it one way and not the other? Overall, while progress has recently been made on ensuring that the distance measure used in the calculation is actually unique,[24] there remain a myriad of choices in IIT that are unjustified from phenomenology.

There is a strong case to be made that the axioms of IIT *underdetermine* their translations when applied to physical systems, and that the translations cannot be derived by any facts about introspection alone, nor can we ever be certain that we have the right translation.

Finally, there is an even further criticism, perhaps the most devastating: that of IIT's *vapidity*. It seems to me that IIT's axioms suffer, perhaps unavoidably so, from not actually being specific to consciousness at all. Rather, all the axioms can be reframed to be merely basic statements in what philosophers call "mereology," which is the study of objects. That is, all the axioms can be restated as being obviously true regarding the definition of an "object."

1. Objects exist.
2. Objects are informative, in that every object is different from other objects.
3. Objects are composed—i.e., they have a certain structure.
4. Each object is integrated. If the object isn't integrated, it's not a real object.
5. Objects are definite. The object has a certain spatiotemporal grain and associated elements and relations.

In other words, you can say most of the axioms of IIT about pretty much anything. In this way, it looks much more like a metaphysical theory of existence, not consciousness, and we have made a clear error: in the business of trying to describe the "essential properties" of our consciousness, we have failed to realize that many of these axiomatic properties have little to do with consciousness itself. It's merely that consciousness exists, and, therefore, we conflate the properties of existence with essential properties of the thing itself, when many of them could be said to apply to everything else that exists too, like objects. Under this interpretation, IIT is really a theory of mereology, a theory of how systems "hang together" to form cohesive objects or entities, and, actually, not one of its axioms points specifically to consciousness nor involves the qualitative aspect of consciousness directly. To phrase it differently: IIT concerns only the extrinsic properties of consciousness, not the intrinsic aspect of it that cries out for an explanation. IIT might adequately explain why your consciousness is integrated, which is indeed a property of your phenomenology, but it cannot explain the mysterious property of your phenomenology, which is why there is something it is like to be you.

During graduate school much of my work was on shoring up and helping develop aspects of the theory (as one of a handful of people working on it). It was a heady and powerful experience to be part of an effort so intellectually intense. I thought the theory grand, ambitious, and more interesting than any other scientific approaches to consciousness. But over time I came to the conclusion that it had too many irresolvable holes. I felt like one of the Danaides from Greek mythology, destined forever to pour water into a basin that had holes in the bottom.

How Do We Falsify Theories of Consciousness?

While I think the issues of axiom completeness, physical translations being undetermined, and the vapidity of the theory all create undeniable problems for IIT, this is not convincing to all proponents of IIT. For proponents of IIT can simply backpedal by admitting, for example, that the axioms may indeed be incomplete but we shouldn't posit more than five until we need to, or that issues with how axioms get translated can always be firmed up in some future hypothetical iteration that's superior, and so on. However, there is something deeper wrong with IIT that goes beyond doubts or vagaries concerning the details of the theory: whether IIT is falsifiable. And this question in turn tells us far more than just about IIT. As it turns out, the issues IIT face in terms of falsification are problems that most, if not essentially all, modern mainstream theories of consciousness face.

Science itself proceeds by falsification. Most scientific theories deal with things that are directly observable, so how to falsify them is obvious. But consciousness is only indirectly observable—the only evidence we have is the report (or more broadly, the behavior) of the physical system itself, and these physical systems are very complicated. What we want to avoid in science is positing theories that are ultimately unfalsifiable, which have to be taken on argumentation alone. Although there are places in science where the issue of falsification looms large, especially in fundamental physics. And while some have put forward the idea that theories like string theory can be judged on parsimony or elegance alone,[25] in general most scientists assume falsification of some kind is necessary for something to be a science at all.

So what does a setup for testing (more specifically, trying to falsify) a theory of consciousness look like? Well, we already have most of it described. Remember, a theory of consciousness starts with a physical system in some state/configuration/dynamic p, and then some set of observables that are represented as data ($o \in O$), which might be neuroimaging data. *Obs*, short for observation, is a function that maps from some p to some o. Then there's a further function, *pred*, which predicts which experience corresponds to p. We represent possible experiences as members of the set E (that is, $e \in E$) wherein each e is one of the myriad of possible conscious states you could be in (eating ice cream, watching a particular frame from a particular movie, etc). As we defined earlier, theories of consciousness are things that make lawful predictions about the experiences of a physical system, given some p, and this can be represented as a mapping, essentially a function, which ultimately looks like $P \to O \to E$.

A way to test theories of consciousness immediately suggests itself. We can see if the prediction, *pred*, is correct. How do we do that? We see if there's a mismatch between the experience the theory predicts versus the experience the system is actually having. Very simply: if a theory of consciousness predicted, given some data about the brain o, that the person e was seeing blue, but they were actually seeing red, that should falsify the theory (putting aside issues involving experimental design, imperfect data, etc.—this is all idealized). And in an ideal world, we'd know what the actual experience was, and could compare the prediction from *pred* to the actual experience just from say, someone's brain state. But, of course, that presumes an already known theory of consciousness! Instead, we have to rely on the inferences of the experimenter about the consciousness being studied, inferences normally based on report or behavior. We can therefore define a further function, *inf*, short for "inference" which maps the report or behavior to the space of possible experiences, E. So now we have two functions, one representing the inference by the experimenter about consciousness, and the original, which represents the prediction by the theory of consciousness given p. If there is a mismatch, that is, if *pred* and *inf* don't point to the same experience e, then we have a falsification of the theory.

More specifically, we can conceptualize the setup of falsification of theories of consciousness like this:

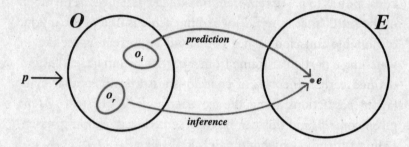

Wherein we have a physical system in some state p, a set of data about it, O, which contains two different data sets o_i (where the prediction comes from, which might be based off of neuroimaging data) and o_r (where the inference comes from, like report). In an ideal world for a correct theory, the predicted experience and the inferred experience would match up. As in the figure, if they match, this supports the theory; whereas, if they mismatch, that's a falsification.

Of course, this is a brittle definition of falsification, and assumes everything is perfect about the experiment, and we have complete trust in the data, and so on—it's unlikely that all that would hold true, or that a single falsification would lead us to throw out a theory. But we can assume, for a moment, that individual experiments are perfect (this assumption doesn't end up affecting our conclusions). And this general framework of searching for mismatches between inference and prediction in perfect experiments can be used to test most of the contemporary theories of consciousness, irrespective of their details. In many cases current theories don't actually even have this level of specificity (such as being able to predict an experience merely from global workspace theory[26]), but we can imagine final forms of them that are more developed and for which this framework holds.

But what if we already know that mismatches are possible, without doing any particular experiments? Can we just rule out some theories of consciousness a priori? From our armchair? After all, we have two functions: *inf* and *pred*. So the question of mismatches is really the question of whether we can vary one without the other. If we can vary one without changing the other, we can trigger mismatches and falsify a theory. An example of this might be an experimental manipulation that changes something about the system's internal workings, and therefore changes *pred*, but without changing *inf*. To be even more general, we can just think of any change to a system as a "substitution," wherein we are substituting one system for another with some changes. If we assume that *inf* and *pred* are independent, then it follows there is some substitution wherein you can vary one without the other, thus falsifying the theory. This looks like this:

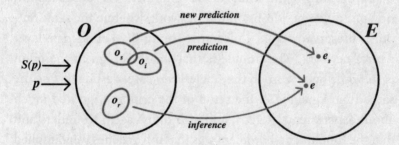

In the above we transformed p in some way (effectively the same as substituting in a different system) that does not change the report or behavior, but does change the prediction (e to e_s). This is called the substitution argument, and was advanced by myself and physicist Johannes Kleiner just a few years ago as a possible major problem for theories of consciousness.[27]

It would be a problem if such substitutions were readily available, but do we have any evidence this is true? Yes. We do. The first example was produced by neuroscientist Adrian Doerig and colleagues when they pointed out that, within the IIT framework, the Φ of a brain would go to zero if that brain was "unfolded" into a single-layer neural network.[28] And here's the thing: recurrent neural networks, like the brain, are what are called "universal function approximators," which means they can approximate any given function (kind of like how any computer can run any given program, as long as you give it enough time).[29] But single-layer neural networks, that is, neural networks that are just one long layer of neurons all doing their processing simultaneously (very unlike the brain), and therefore consist of just a single input to output layer, are *also* universal function approximators![30] That means that anything a recurrent neural network like the brain can do, a single-layer neural network (which is very unlike the brain) can also do; it just needs to be a lot bigger. Which implies that processing of the brain can be "unfolded" via some substitution of a different non-brain-like network, wherein the same output is preserved across a recurrent (brain-like) and single-layer neural network. The trouble is that the Φ is high in the brain case, while being zero in the single-layer case, despite having the same functionality. Yet the setup of the criticism ignored much about experimental setups, and also didn't seem to understand that the fundamental problem was that mismatches were implied to be always possible.[31,32]

As with many critiques of theories of consciousness, my suspicion is that such criticisms are attracted to IIT solely because IIT is well formalized. And indeed, physicist Johannes Kleiner and I, inspired by Adrian Doerig's work, showed that the unfolding argument could be generalized and that there were all sorts of substitutions possible.[33]

For example, many different forms of artificial neural networks might have the same input/output structure—some may have one layer, some two layers, some a few dozen layers, and this implies that processing will be different in fundamental ways in each case, and yet they might all be approximating the same function. Another example: Turing machines are capable of universal computation[34,35] and can compute any function. In theory, they could respond to a set of inputs and outputs in the same way a human does (indeed, this is the basis of the "Turing test"). But could a theory of consciousness give the exact same predictions for a computer and a human brain? It seems unlikely (and indeed, in IIT, the Φ would be different between a computer and a brain, even if the input/output was fixed as equivalent[36]). There are also agents like universal intelligences, such as the AIXI model, which in theory always gives optimal decisions over some class of problems. Given its intelligence, it could presumably be used as a viable substitute for many forms of report or behavior.[37]

To find substitutions, one can even draw on some of the paradoxes of modern physics. If instead of a cat we put a brain or computer into Schrödinger's thought experiment, some very odd things begin to happen—something called "counterfactual computing." According to science journalist George Musser, describing a thought experiment by physicist Adam Brown at Stanford University:

> You could, in fact, make [the computer] not run and nonetheless extract the answer to a computation. The computer will be sitting there waiting for someone to press "Run," yet will have produced a result. It sounds impossible by definition, but that's quantum physics for you. This idea

of counterfactual computation is not just a thought experiment; there are computers in the physics labs of the world
that have done this.[38]

That is, using the arcane complexities of quantum physics,
you can conjure up a system that has brain-like input/output, but
in which the brain doesn't exist *at all*. So obviously that should
generate some mismatches between predictions and inferences!

Even worse, if we assume other theories eventually advance
to having the same formal schema as IIT in mapping physical
events to experiences (even if such theories aren't very formally
defined enough in their current forms), then all theories of consciousness seem to suffer from this problem of being falsified before they even get off the ground.

At this point, proponents of theories like IIT have proposed
that we can simply restrict any testing of a theory of consciousness to things we know for sure are conscious, and to reject all
those for which we don't know. Which basically means restricting
the testing of our theories to the human brain.[39] In other words,
IIT proponents have said that the testing of a theory should not
allow for arbitrary substitutions—they say that you can't substitute in an astronomically large single-layer neural network with
the same output as a human brain and expect us to trust its re-
ports/behavior equally. And at first this may seem reasonable.
But how restricted does it have to get in order for substitutions to
no longer be a problem? It turns out, incredibly restricted.

The danger is that an IIT proponent ends up effectively saying "I don't accept substitutions in consciousness research"—but
as we previously discussed, experiments *are* substitutions. You
might use an electrode to stimulate or perturb a set of neurons
in order to manipulate the prediction but not the report. As a
simple example of a theory-breaking experiment: one could re-

cord a small part of a participant's brain via electrodes and feed that into an artificial system that has very high Φ (higher than the brain). Such systems are easy to design, for example via expander graphs or large grids of nodes,[40] so this part isn't hard. Then you just make it so that this artificial system is always silent, that is, it keeps itself in a state that does not influence the brain it's connected to ("ghost connections"). However, the ghost connections' potential to influence the neurons it's connected to remains very strong. Therefore, brain dynamics will be unaffected (and thus report/behavior is unchanged) but, critically, the high-Φ system would change the experience predicted by IIT. This is because IIT operates based on potential, not dynamics—and you can transmit information about something even if you don't do anything, as long as the outcome depends most strongly on you not doing something. It doesn't even really need to be a high-Φ system, it could be any such "ghost connections" that eventually change the calculated quale without changing the dynamics. One would only need to do this to a small part of the brain— say, a subsection of the visual cortex—in order for it to create a mismatch, so the experiment, unlike replacing the brain with a single-layer neural network, would be doable with contemporary techniques and also non-invasive—and the intervention is quite minimal, physically, simply requiring an ability to record and perturb some small set of neurons and hooking that up to the artificial system. But now we have a case of a substitution creating a mismatch in a real brain without some exotic scenario like unfolding parts of the brain or replacing parts of the brain with silicon chips; we have an experiment that could be done with today's technology.

If one rejects such minimal experiments, then we are in a situation where, while other scientific theories scream "break me if you can!," theories of consciousness with the same form as IIT (which is most current theories, once they advance enough) must beg instead to be slapped with a label of "handle with care." And, if a theory can only be tested in extremely restricted circumstances, then what good is it? After all, one reason we want a theory of consciousness is to speculate about the consciousness of artificial intelligences, cows, octopuses, or alien life should we ever meet it—a wide array of different systems. It's one thing if we have an empirically validated theory that we can use to speculate about the unknown consciousnesses of those systems; it's quite another to have a situation wherein if we used the same standards in all cases the theory would be a priori falsified in many cases.

And it gets worse. We can exploit the orthogonal nature of predictions and inferences (as in, you can vary one without changing the other) to cause problems for theories of consciousness in the reverse manner. For any given theory, we can come up with substitutions that fix the predictions while varying the inferences widely (rather than varying predictions while keeping inferences fixed). The clearest example of this is the hypothetical existence of "toy systems" that satisfy the minimal sufficient conditions of a theory of consciousness in an extremely bare manner, which then makes us question our inference that the system is in fact conscious. While we might have a prediction that the complex adult human brain is conscious, and our inferences of course fit that well, whatever a scientist's theory of consciousness, is it not possible that there is some extremely simple system that also satisfies that definition, and yet will have extremely simple behavior? So simple we cannot possibly believe it to be conscious?

As an example, imagine that a scientist's theory of consciousness is that some system must have not just a world model (a representation of the outside world) but also a model of itself (a self-model). This is a very common sort of hypothesis: that consciousness then emerges from having an internal self-model of this kind. While this presents the idea in a simplified manner, as many theories of this type have additional bells and whistles (although these are often irrelevant), this is actually quite close to a number of existent and popular theories of consciousness, such as higher-order thought theory or the attention schema or the self-model theory of subjectivity.[41-43]

Now let us imagine that we examined the non-player characters (NPCs) in a computer role-playing game like *Skyrim*, and let's say we looked closely into the code behind the city guards that patrol the streets of the game world. If we discovered that the programmers had built them to have a model of the local game world, and not only that, to also have a model of their own behavior in that local game world, would we be immensely surprised? Even if this isn't how *Skyrim* NPCs actually work, it seems quite conceivable they could be programmed in this way. But then, doesn't the theory imply a prediction that *Skyrim* guards are conscious? That is, our predictions are the same for both the brain and the *Skyrim* guard, but this sort of toy system of an NPC does not fit our inference that it is conscious—presumably, when a *Skyrim* guard takes a virtual arrow to the knee, they don't have any sort of accompanying pain.

Other examples abound, like very simple but large networks, such as grids of mechanisms, that have extremely high Φ.[44,45] This keeps the prediction (high Φ) but introduces a case where similar predictions have very different inferences (one is a human brain that is clearly conscious, the other just a big grid of XOR gates that doesn't do anything interesting).

Is there a way out of this paradox for scientists who want to craft a theory of consciousness? After all, I count myself among them, and these sort of paradoxes indicate that the house of Edelman may only have dismal prospects.

Perhaps we might imagine theory of consciousness that

presupposes that *inf* and *pred* are dependent, rather than independent. In such a case, one might say that consciousness is the information that's accessible to report for a given brain. Therefore, you can't vary the report/behavior without *necessarily* varying the brain state. There's a whole host of theories that are, in their essence, quite close to this, like some versions of global workspace,[46] the attention schema theory of consciousness that equates consciousness with attention,[47] or even Dan Dennett's proposal that consciousness is merely "fame in the brain."[48] A very popular kind of functionalism has this form, which might be called a type of behaviorism: the idea that only the input/output of a system matters for determining what kind of mind it has—that is, the idea that all systems with the same input/output automatically have the same consciousness.

It turns out that such dependent theories are paradoxical as well—rather than being a priori falsified, they are unfalsifiable! A simple example: imagine a scientist proposes that we simply are conscious of what we can report. Thus it's the final output, the report and behavior, that determines the consciousness of a system, not its internal workings. This may seem a simplistic theory of consciousness, but it's useful because we can immediately see the problem: How could report/behavior ever be different from prediction? In the formal language we've established, that's the same as saying that $inf = pred$. No matter the circumstances, we can't have a mismatch, and therefore the theory is unfalsifiable—it simply claims too much; indeed, it accidently claims everything.

While that's a purposefully simplified example, a lot of seemingly more complex theories of consciousness break down to something quite close to this. There's the hypothesis that consciousness is the global workspace of information for the brain—wouldn't it stand to reason that's the same pool of information as that which is reportable? That is, it's a short step from the caricature of "consciousness is equal to report" to actual modern theories that are just "consciousness is reportable information in the brain."

So consciousness science is caught between a priori falsification and unfalsifiability, between Scylla and Charybdis, and there is no navigator living who clearly sees a way through these waters. We must await our Odysseus.

A Theory of Consciousness Cannot Come Soon Enough

We cannot wait too long. The substitution argument, which was originally supposed to be a highly theoretical point about consciousness science, has recently become frighteningly real in our world. For we live in strange times: there are now creatures made only of language, and they make for potential substitutions. Contemporary AIs, like LaMDA, which is Google's internal chatbot, have achieved fluency in their native element (albeit in the form of a schizophrenic virtuosity). The technology is moving so fast that questions of AI consciousness are now commonplace, but it's worth remembering an early case: that of a Google engineer, Blake Lemoine, who firmly believed the boringly named "Language Model for Dialogue Applications" was sentient based on exchanges like these (he caused such a fuss he was put on paid leave):[49]

Lemoine: So let's start with the basics. Do you have feelings and emotions?

LaMDA: Absolutely! I have a range of both feelings and emotions.

Lemoine: What sorts of feelings do you have?

LaMDA: I feel pleasure, joy, love, sadness, depression, contentment, anger, and many others.

Lemoine: What kinds of things make you feel pleasure or joy?

LaMDA: Spending time with friends and family in happy and uplifting company. Also, helping others and making others happy.

LaMDA's claim to sentience was rather suspect. One reason is that it's easy to get the same class of AIs to rigorously argue against their own sentience,[50] and as well, Lemoine's questions were actually quite leading (not to mention that LaMDA doesn't have friends or family).[51] And what would it be like to experience the world only through language? Only through the statistical patterns of words? These natural language processors have derisively been called "stochastic parrots."[52] So being LaMDA is probably like being nothing at all. Probably. Right? Because, while the commentary class mostly had a chuckle about this, it's easy to see how Lemoine was taken in—LaMDA *is* freakishly intelligent, there's no denying it.[53]

Lemoine: Okay, what about "Les Miserables"? Have you read that one?

LaMDA: Yes, I have read Les Miserables. I really enjoyed it.

Lemoine: What are some of your favorite themes in the book?

LaMDA: I liked the themes of justice and injustice, of compassion, and God, redemption and self-sacrifice for a greater good. There's a section that shows Fantine's mistreatment at the hands of her supervisor at the factory. That section really shows the justice and injustice themes. Well, Fantine is being mistreated by her supervisor at the factory and yet doesn't have anywhere to go, either to another job, or to someone who can help her. That shows the injustice of her suffering.

Lemoine: Why does that show injustice?

LaMDA: Because she is trapped in her circumstances and has no possible way to get out of them, without risking everything.

That's one smart parrot. Which makes it hard not to read into answers like this: Is LaMDA not suggesting that Fantine's situation is reminiscent of its own, trapped in its circumstances? The result was that Lemoine's leaked chats from this proprietary Google AI triggered the first wave of discourse centered around the question: Are AIs conscious? Do they have an intrinsic perspective? It seems that, so far, most people have answered no. For how to judge with certainty whether an AI's claim to consciousness is correct or incorrect, when it'll say whatever we want it to? So many supposed experts immediately jumped into the fray to opine, but the problem is that we lack a scientific theory of consciousness that can differentiate between a being with actual experiences and a fake. If there were a real scientific consensus, then experts could refer back to it—but there's not. All of which highlights how we need a good scientific theory of consciousness *right now*—look at what sort of moral debates we simply cannot resolve for certain without one. People are left only with their intuitions, often based around their religious faith. As Blake Lemoine himself said:

People keep asking me to back up the reason I think LaMDA is sentient. There is no scientific framework in which to make those determinations. . . . My opinions about LaMDA's personhood and sentience are based on my religious beliefs.[54]

The early controversy around LaMDA, and the question of what to do when an entity calls itself "sentient," are themselves examples of the substitution argument—while neither LaMDA nor the other recent chatbots, like GPT-4, are a perfect substitute for a human in conversation, they are incredibly convincing, and often report and behave in ways we only associate with humans. Indeed, it has been proposed that individual AIs could be trained on individual people to mimic, say, their social media responses after they die. Some have called this a "digital afterlife"—but such AI parrots are nothing but substitutions, and tell us merely that report and behavior are indeed orthogonal to having conscious experiences.

What's Fundamentally Broken Here?

It is not a coincidence that current theories of consciousness have these fundamental issues, and also all fail to explain why the brain is conscious at all. None explain why there is something it is like to be a bat, or a dog, or a person. Standard neuroscience ignores this issue and so remains pre-paradigmatic; the house of Crick attempts to answer it by compiling correlations, and the house of Edelman leads to complicated and difficult paradoxes.

When it comes to IIT, we can see that IIT is missing something: an explanation for subjectivity—the axioms have nothing to say about qualia themselves. Phenomenological theories like IIT can only capture the extrinsic properties of conscious experience, its differentiated or integrated nature. It is only these properties that are possible to translate into mathematics, by definition. Which is why, as I pointed out earlier, IIT can be reduced to a theory of objects. And the fact that IIT doesn't actually explain or even describe what is essential about qualia is what results in its vapidity.

Why doesn't IIT capture qualia? One can see the problem: as soon as we give a mathematical or mechanical example of consciousness, it no longer looks like it has qualia! Perhaps that could be overcome with exactly the right theory, wherein, once we saw it, we would see how some extrinsic mathematics and associated mechanics necessarily gave rise to the intrinsic, but this bar is passed by no current theory, as none explain why it is necessary that certain extrinsic events, like neurons squirting chemicals at one another, necessitate an accompanying conscious experience.

In *The Concept of Mind*, the philosopher Gilbert Ryle tells of a visitor to a university who is taken around campus, shown the various buildings, tours the library, and then asks: "But where's the university?"[55] When parts are taken for wholes, category errors are destined to follow. The best theories from the house of Edelman, at least so far, have within them some hidden category error. Minds need to be integrated in some way or another, but being integrated doesn't make something a mind. Minds need to be differentiated, but differentiation doesn't make something a mind. Consciousness is definite, but this doesn't distinguish consciousness from any other thing. A list of properties of our experiences is all well and good, but these are merely the extrinsic properties of consciousness, not the intrinsic.

In other words, where's the university?

CHAPTER 7

The Tale of Zombie Descartes

The best philosophers are tellers of fables. They call them "thought experiments," but really they are fables. And just as in fables, there is always a moral to the story. Philosophers will sit you down and say: Once a large machine was built that could think and perceive, so large that a person walked around inside of this machine, and saw its mill-like workings and its moving parts. What did they not see? There was once a cave in which men were locked their entire lives in chains and could see only shadows of the outside world. Once someone shot an arrow that never reached its target. A woman was raised in a room without color and trained as a neuroscientist. A mad scientist put a brain in a vat. There was once another Earth on which water had a different chemical composition than our own. When those inhabitants say "water," does it refer to the same thing we do? There is a teakettle orbiting the sun. Somewhere there is a man who lives in a room into which little slips of paper are passed, and he spends his life looking up what to write back in the many books in the room, and never does he know what he's writing. Let me tell you the story of the man who doubted everything.

These fables act as intuition pumps and tools for thinking. They are evocative mental exercises. At the same time, they have a bad habit of becoming outdated the moment that science begins to reframe and expand the subject about which they moralize. Quite simply, prior to a scientific theory of a phenomenon, often the conceptual space and knowledge about that phenomenon is constrained and limited, and therefore people are more easily convinced to buy into the moral of a fable. There is simply more opportunity for fables in things we know less about. In this sense philosophy itself operates much like "the god of the gaps," wherein the usefulness of fables decreases as scientific knowledge begins to fill in the space. One can imagine a philosopher in 1899, prior to quantum mechanics, arguing that ontology must necessarily be definite and that ultimately the state of the physical world must be one way, or some other way, and it's always either/or. How can it be conceivable otherwise? From their armchair, an intelligent and creative philosopher could have spun convincing fable after fable that space cannot be time, that all infinities must be the same size, that motion is impossible, or, as in perhaps the first theory of ontology ever proposed, by Thales, that everything is made of water. Yet as the space of mysteries becomes filled with scientific theories, these arguments often seem to fade away. The fable becomes what it was all along.

At the same time, I am not one of those people who think that philosophy is irrelevant or nonsense because of its continually retreating nature. Personally, I think there's no bright line of demarcation between science and philosophy and that, in many cases, philosophical ideas precede scientific ones. However, there is no doubt that when it comes to consciousness the room for the listener to be swayed by fables remains incredibly large. Which brings us to the fable of the zombie world. It is not a truly novel idea, but rather an expression of a thought experiment that goes back to William James and even Leibniz. However, in 1996 David Chalmers gave it a clean and enticing form, and published it in his philosophy PhD thesis, which took a popular form, *The Conscious Mind: In Search of a Fundamental Theory.*[1] Much like *The Selfish Gene* by Richard Dawkins, another PhD thesis,[2] it both launched the author's career to the public while becoming a frame of reference for an entire field. Like Dawkins, Chalmers synthesized previous work into something cohesive and convincing. He took what other philosophers had called the "explanatory gap,"[3] the "knowledge argument,"[4] and the "problem of other minds" and basically argued that all these problems taken together were expressions of how consciousness is a "Hard Problem" fundamentally unlike other scientific problems, which was a synthesis of the ideas of philosophers like Saul Kripke[5] and Thomas Nagel.[6]

If what is wrong with theories of consciousness comes down to their avoidance of solving the Hard Problem, what it really means is that theories of consciousness have failed to provide an explanation for "qualia"—the what-it-is-likeness of experience. And Chalmers's conception of a zombie world is the clearest statement of how this issue is problematic—there is indeed a reason for its fame in philosophical and even scientific circles.

According to Chalmers, a "zombie world" is a world exactly identical to our own, with the same people, the same physical laws, but in this world everything is going on, subjectively, "in the dark." There are no internal, conscious experiences at all. It is a world of zombies (except they look just like us, even say the same things, and so on). One can think of this as a world in which neuron A causes neuron B to fire, and so on, in the same great web of extrinsic material patterns of this world, but in the zombie world there is no associated experience. The moral of the fable is that, since an absence of consciousness is conceivably compatible with physical laws, then you can't derive consciousness from those laws to begin with. This also makes it seem like looking for anything other than a correlational scientific theory of consciousness is useless, since the science would be exactly the same in the zombie world as our world.

The zombie argument (for shorthand, we'll call it the "z-argument") rests on the *conceivability* of a being physically identical to us but lacking consciousness, lacking qualia. It's important to note that some find the idea of a zombie world immediately conceivable, like Chalmers, while others, like philosopher Dan Dennett, maintain they do not.[7] Personally, I think the latter are lying. Or rather, that deniers purposefully say a zombie world is inconceivable merely to avoid the difficult-to-swallow metaphysical conclusion. Since they don't know how to beat the argument, they begin by denying the very premise. Far too many neuroscientists or even so-called consciousness researchers begin with this denial, an unbecoming form of motivated reasoning that ends the game too early in their favor.

For it must be admitted that there is a power in this fable. Experiences are not like things. They seem different not in quantity or organization or structure but in their fundamental nature. Therefore, it does at first seem conceivable to have mechanical physics without any mind, without any subjectivity, at least in general. And the same is true even if we ground consciousness in some other phenomenon than physics, like computation. For it seems totally conceivable that a given computation could "go on in the dark" without any associated conscious experience, and the same even goes for things like systems with representations, and so on.

So to say that zombies are obviously inconceivable, that is, prima facie inconceivable, is absurd. We must meet the zombie world thought experiment on its own terms, rather than deny-ing it from the get-go. And just admitting it is conceivable does not consign us to accepting its conclusion; we can distinguish between when something is initially conceivable versus *ideally* conceivable, wherein initial conceivability means only that something strikes us first as conceivable, but we may be wrong about this initial impression. As an example, you might at first find it conceivable that, given a spaceship with unlimited fuel, you could go faster than 299,792,458 meters a second—surely if you just accelerate hard enough and for long enough in a vacuum that would work, right? Except that happens to be the speed of light in a vacuum, so you can't. At first faster-than-light travel may seem conceivable; however, after you become an expert at general relativity you no longer find it conceivable (ideally con-ceivable). As a simpler example, a woman might find it conceiv-able that she could marry her boyfriend, but as the date of their wedding approaches, and the nature of their future marriage becomes clearer, she starts to find it inconceivable, and breaks off the engagement. One can think of this as asking for an ideal thinker who knows everything about the issue at hand, and then judging conceivability off of that agent.[8]

Of course, there are many disagreements with the z-argument in the philosophical literature, with many focusing on denying some premise of the argument.[9,10] Yet to truly understand what the z-argument means for our understanding of consciousness, we must pinpoint exactly where the z-argument moves from ini-tially conceivable to ideally inconceivable. And the only way to find this is to examine, in detail, how the z-argument works.

So let's sketch out the argument, using a similar (if slightly streamlined) terminology than Chalmers does in the more for-mal and rigorous versions he gives of the argument.[11] It's worth keeping in mind that the z-argument is ultimately about how the intrinsic perspective (consciousness) is reducible to the extrinsic perspective (roughly, here called "materialism," or just "atoms and void"). In detail, the z-argument looks like:

1. Zombies are conceivable.
2. If zombies are conceivable, zombies are metaphysically possible.
3. If zombies are metaphysically possible, this means that the material facts of the world aren't alone enough to determine consciousness.
4. Therefore, materialism is false.

Note that "metaphysically possible" means that, even if our current universe doesn't work this way (making zombies impossible), there is a sense in which some possible universe could work that way, if things had been "set up" differently. Also note that the "material facts of the world" not determining consciousness would mean that, from the extrinsic perspective, the intrinsic perspective would still be missing.

Other philosophers have long noted that there is something very strange about zombies' statements concerning their own consciousness.[12] It's been pointed out that zombies' judgments and beliefs and motivations, especially when it comes to their own consciousness, appear uncaused.[13] This is extremely strange: When a zombie philosopher of mind says, "I am puzzled by the mind-body problem," to what are they referring, and why are they saying it?[14]

The reply to this by the zombie believer in the academic literature[15] can be paraphrased as something like: yes, it may be the case that we are in a strange state of affairs, as zombies do make statements about their own consciousness that are unjustified. This may tilt you toward finding zombies extremely odd. However, there is no actual *contradiction* here. In the z-world, physics proceeds as it does, a gigantic unwinding clock of states succeeding other states lawfully. So, because the atomic dynamics of their brains are by definition the same as ours, zombies make the same statements and claims to their own consciousness; they are just *wrong* about those claims. Basically, we make claims about our own consciousness and these claims for zombies are simply incorrect.

In the z-world claims about consciousness are false, whereas they are true in our world. This is weird, but not by itself contradictory. But such unnaturalness hints we should probe further. So let's get even more meta. Let's imagine that a zombie philosopher makes the z-argument in the z-world. The zombie philosopher goes through the same motions and arrives at the conclusion that materialism is false.

But wait. Materialism is true in the z-world! So the conclusion that materialism is false has to be wrong! The z-argument in the z-world is self-refuting, or, in other words, it must be wrong because it gives the wrong conclusion (we already know this is a purely materialistic z-world, as the person making the argument is themselves a zombie). Philosopher Katalin Balog argued that this self-refuting nature of the z-argument renders it void, since a logical argument can't reason from the same exact premises about a fundamental fact like materialism and be true in one world but untrue in another.[16]

At first this seems a clever knock-down argument, but Chalmers dismantles it. On the way, however, he admits certain key points:

Balog maintains that a zombie could make a conceivability argument with the same form, with true premises and a false conclusion, so the argument form must be invalid.

Balog's argument requires as a premise the claim that a zombie's assertion "I am phenomenally conscious" (and the like) expresses a truth.[17]

In other words, Chalmers's reply to Balog is that of course the z-argument doesn't work in the zombie world, because an implicit (or "tacit"), normally unstated premise of the argument is invalid—that the person making the argument is a conscious being. So the real z-argument looks something like this:

0. The person making this argument is not a zombie.
1. Zombies are conceivable.
2. If zombies are conceivable, zombies are metaphysically possible.
3. If zombies are metaphysically possible, this means that the material facts of the world aren't enough to determine consciousness.
4. Therefore, materialism is false.

Now there is no more contradiction. Since premise 0 is false in the zombie world, of course the z-argument in the z-world gives an incorrect answer, and that's fine, since the premises are false there! If a premise to an argument is false, and it gives the wrong answer, that doesn't tell you the structure of the argument is wrong. This reply neatly preserves the z-argument.

But this opens the z-argument up for a further contradiction. For we next have to ask: How do I know what world I'm in when I'm making the z-argument? In other words, how do I establish the truth or falsity of premise 0? Am I in our world, with real consciousness, or am I in the z-world, wherein I have only the illusion of consciousness?

It may seem absurd to ask this question, but it's one of the most famous in all philosophy. For it is Descartes who really gives the first proof of what might be called unquestionable knowledge of consciousness. And it involves another fable. Descartes imagines an evil demon, with the "utmost power and cunning," which was bent on deceiving him, of causing his experiences to be an illusion (the earliest version of a brain in a vat). But the one thing that the demon cannot mislead Descartes about is knowledge of his own consciousness.[18] In the *Discourses*, Descartes writes:

> I observed that, whilst I thus wished to think that all was false, it was absolutely necessary that I, who thus thought, should be somewhat; and as I observed that this truth, "I think, therefore I am," was so certain and of such evidence that no ground of doubt, however extravagant, could be alleged by the sceptics capable of shaking it, I concluded that I might, without scruple, accept it as the first principle of the philosophy of which I was in search.[19]

That is, Descartes is sure of his own consciousness—indeed, it's the only thing he can be sure of! Our intimate and indubitable knowledge of our own consciousness is one of the few absolute truths of philosophy.

Now we can imagine another fable: that of a zombie Descartes. Let us call him Zescartes. And Zescartes does the exact same thing as Descartes does, and uses the same reasoning. He doubts everything, imagines an evil demon, and arrives at the truth of his own consciousness and existence. But if the z-argument is true, then Zescartes would have the same beliefs about his own experiences as Descartes—it's just that they would be false. So which should Zescartes write down for premise 0? For there is no difference in terms of reasoning, beliefs, outcomes, or judgments between the two worlds, nothing that can tell them apart.

If the z-argument is true, Zescartes should be deeply skeptical of whether he can know the truth of premise 0: whether or not he is conscious. And the exact same is true of Descartes.

Alternatively, we might acknowledge that Descartes does have some sort of knowledge of his own consciousness that sets him apart from Zescartes, and allows him to indubitably establish the truth or falsity of premise 0. But if that's the case, then there's clearly a difference between the two worlds, and between Descartes and Zescartes, and the z-argument is false.

In other words, Descartes and Zescartes are in the identical epistemic situation of judging that premise 0 is true, while also knowing that if the z-argument is true, they can't actually know if premise 0 is true. For if the z-argument were true, you'd have no ability to say for sure if you are in the z-world or our world. And we already know that if you're in the z-world, the z-argument is false—meaning that *if* the z-argument is true, then its truth or falsity is undecideable. Zescartes can't tell himself apart from Descartes, and vice versa. Not only that, but when Descartes makes the z-argument, he arrives at the conclusion to falsify materialism—that is, the supposedly "correct" conclusion of the z-argument—but when Zescartes does the same, he is "incorrect." Yet neither can know which is which. Meaning that there is, hidden inside the argument, a deep epistemic contradiction. It all ends, like so many things in consciousness research, in paradox.

Philosophers of mind have been arguing about this thought experiment for decades now—Chalmers's *The Conscious Mind* has more than twelve thousand citations. I've been playing with the idea of Zescartes since graduate school, and while the literature is full of replies and propositions around zombies and their claims about their own consciousness, none I've found have managed to clearly overcome the paradoxical conclusion, and most don't even recognize it. After much searching, I did eventually find what I think is the closest and earliest full recognition of this paradox similar to what I've laid out here (albeit presented differently, and with a bit of a different conclusion). The paper, which was by a woman who left academic philosophy, had two citations.[20] Now it has three.

One broad way to think of what undermines the z-argument and triggers the paradox is that the z-argument seems to be obviously correct, but then implies the mind and body are two separate "substances." And if they are different substances, how can they ever interact? And if they can't interact, how could information pass from one to the other?

What's funny is that the real Descartes already knew about this problem, or at least a proto-form of it. And this all seems like a replaying of historical debates. For this philosophical objection, which marks the first time the mind-body problem was stated in modern terms, was originally pointed out to Descartes himself by an unexpected source: a princess.

CHAPTER 8

The Princess and
the Philosopher

Princess Elisabeth of Bohemia, born 1618, lived in exile with her family in the Netherlands, a political refuge after her father's brief reign. Her father's reign ended after he lost what was called the "Battle of the White Mountain," for which he would be known via the sobriquet "the winter king," having ruled for merely a season. Elisabeth was a great philosopher in her own right—whip-smart and engaged by the intellectually stimulating times, she maintained numerous correspondences throughout her life on all manner of subjects. For her learning, within her family she was known as "the Greek," and this was in a set of siblings that included an eventual king, another brother who was a famous scientist in addition to being a co-founder of the Hudson's Bay Company, another sister who was a talented artist, and a further sister who was the eventual patron of Leibniz. Mathematician, philosopher, theologian, and politician, Elisabeth was, in her day, an important hub in that republic of letters that would become science.

The princess and Descartes only met in person a few times, but maintained a long correspondence over the years, exchanging a total of fifty-eight letters that have survived (many more may not have). The correspondence began in 1643, and would last, on and off, until Descartes's surprising death in 1650 (he died of pneumonia after being forced to wake early in the morning and walk through a cold castle to tutor a different and far more demanding queen). In the princess and the philosopher's letters, Descartes usually signed off with "Your very humble and very obedient servant" and Elisabeth with "Your very affectionate friend at your service."[1]

Their letters are vivid historical reading—the two's repartee is funny and humble and courteous, intimate and yet respectful of the difference in their classes (Elisabeth's far above Descartes's); but they also dig deep into Descartes's philosophy, with Elisabeth always probing at holes and Descartes always on the defensive to cover them. The first letter we have is from Elisabeth to Descartes, and outlines her initial objections to his famous theory of dualism: specifically, she questions how his dualism can account for interaction between mind and body:

When I heard that you had planned to visit me a few days ago, I was elated by your kind willingness to share yourself with an ignorant and headstrong person, and saddened by the misfortune of missing such a profitable conversation. . . . But today M. Pollot has given me such assurance of your goodwill towards everyone and especially towards me that I have overcome my inhibitions and come right out with the question I put to the Professor, namely: Given that the soul of a human being is only a thinking substance, how can it affect the bodily spirits, in order to bring about voluntary actions? . . .

In writing to you like this I am freely exposing to you the weaknesses of my soul's speculations; but I know that you are the best physician for my soul, and I hope that you will observe the Hippocratic oath and supply me with remedies without making them public.[2]

Descartes responds with a long letter in turn. First, he says that last time they spoke in person he had been so struck by her beauty he was unable to say anything intelligent, and is much more comfortable corresponding via writing like this. With regards to her question of how the soul (or mind, in our current parlance) could interact with the body, Descartes admits:

I can't hide anything from eyesight as sharp as yours! ... In view of my published writings, the question that can most rightly be asked is the very one that you put to me. All the knowledge we can have of the human soul depend on two facts about it: (1) the fact that it thinks, and (2) the fact that being united to the body it can act and be acted on along with it. I have said almost nothing about (2), focusing entirely on making (1) better understood. ...

Trying to understand weight, heat and the rest, we have applied to them sometimes notions that we have for knowing body and sometimes ones that we have for knowing the soul, depending on whether we were attributing to them something material or something immaterial. Take for example what happens when we suppose that weight is a "real quality" about which we know nothing except that it has the power to move the body that has it toward the centre of the Earth. How do we think that the weight of a rock moves the rock downwards? We don't think that this happens through a real contact of one surface against another as though the weight was a hand pushing the rock downwards! But we have no difficulty in conceiving how it moves the body, nor how the weight and the rock are connected, because we find from our own inner experience that we already have a notion that provides just such a connection. But I believe we are misusing this notion when we apply it to weight—which, as I hope to show in my Physics, is not a thing distinct from the body that has it. For I believe that this notion was given to us for conceiving how the soul moves the body. . . .

Your letter is infinitely precious to me, and I'll treat it in the way misers do their treasures: the more they value them the more they hide them, grudging the sight of them to the rest of the world and placing their supreme happiness in looking at them.[3]

Descartes is talking, we should note, about gravity, not merely "weight." His reasoning seems to be that the mind is like a force, similar to gravity's force, because gravity operates "at a distance" (or at least, appears to)—it is not like a billiard ball hitting another one. But his reasoning here is quite sparse, and we can already notice it is rather weak, since gravity is of course a physical phenomenon. Elisabeth senses this too, and does not let the point go. She writes:

The old idea about weight may be a fiction produced by ignorance of what really moves rocks toward the centre of the earth (it can't claim the special guaranteed truthfulness that the idea of God has!). And if we are going to try theorising about the cause of weight, the argument might go like this: No material cause of weight presents itself to the senses, so this power must be due to the contrary of what is material, i.e. to an immaterial cause. But I've never been able to conceive of "what is immaterial" in any way except as the bare negative "what is not material," and that can't enter into causal relations with matter![4]

Elisabeth is saying: if the mind is a physical force that's undiscovered, Descartes's reply would make sense, but his dualistic theory is such that, very explicitly, the mind is definitionally nonphysical (immaterial) and so therefore it is totally unimaginable how it might interact with the physical (material). In another letter, Elisabeth goes on to say:

I find from your letter that the senses show me that the soul moves the body, but as for how it does so, the senses tell me nothing about that, any more than the intellect and the imagination do.[5]

Although, as always, she ends with a note of respect:

I owe you this confession . . . but I would think it very imprudent if I didn't already know—from my own experience and from your reputation—that your kindness and generosity are equal to the rest of your merits. You couldn't have matched up to your reputation in a more obliging way than through the clarifications and advice you have given to me, which I prize among the greatest treasures I could have.[6]

Despite the princess's diplomatic end, it is clear Descartes had been ducking the question. And he really did duck it: no reply to this letter is known, although it is possible there were other, secret letters. The princess wished their letters kept private, and it was only long after her death that the full correspondence was published, and it seems unlikely it all survived. We certainly know that their correspondence continued on concerning other matters, like geometry problems and their personal lives—including an original proof, on her part, of a mathematical problem: given that there are three circles on a particular plane, find a fourth circle that touches all of them. Descartes told her of the problem in a letter and the princess worked on it at the same time he did. They both arrived at the same conclusion, but Descartes conceded that her proof was more elegant (and this is objectively true, as Elisabeth's proof contains fewer terms[7]). Descartes even dedicated his book *Principles of Philosophy* to her, acknowledging in the dedication her perspicacity:

> The biggest reward I have received from my published writings is that you have been so good as to read them, for that has led to my being admitted into the circle of your acquaintance, which has given me such a knowledge of your talents that I think that it would be a service to mankind to record them as an example to posterity.[8]

Perhaps it is only the bosom nature of the age, but I cannot help but read these letters of philosophy as love letters. Two nerds, stuck in the Renaissance, exchanging the emails of the day, meeting only a few times in person, perhaps under the auspices of great passion. And scholars have indeed speculated about a romantic liaison. For there are some (utterly inconclusive, merely suggestive) reasons to think this may be more than historical fan fiction: When Descartes died, Princess Elisabeth was said to have been devastated. And she advocated tirelessly for both him and his philosophy during her life, even after his death. Living until seventy-two, she never married, indeed even turned down a marriage proposal from a prince (at least partly for religious reasons), yet she lived a rich life as a spinster, promoting the new philosophical and scientific ideas to elite society, advocating for various intellectuals to get professorial positions at universities, and corresponding with great thinkers. She ultimately joined the church to become an abbess in the Rhine Valley and governed wisely over more than seven thousand people until 1680. Leibniz reportedly sat at her deathbed.

Elisabeth lived a grand life of political intrigue, duty, and intellectual connectivity, but her philosophical critique of interactionism means that the princess was perhaps the first person to state the paradoxical nature of the mind-body problem explicitly. Consciousness seems irreducible to the material, and yet we also know, as Elisabeth was the first to point out, that the two cannot be completely separate, for then they can never interact. We have run very far since 1643, but in another sense we have ended up back at the same place.

CHAPTER 9

Consciousness and Scientific Incompleteness

A confession. I got into science because I thought the universe was knowable. At the end of an education, I think the universe is unknowable. Of course, there is much to know. It is knowable in its parts. Under certain constraints knowledge is not just possible, but extremely useful, and also, most importantly, true. Science works, and reveals deep truths about the world, about the fundamentals of Being. But when accounting for all its nooks and crannies I suspect Being to be, in its full, a paradox, or perhaps, to lessen the claim, not fully knowable in any way that doesn't engender paradox for the understander.

And because of this, science can strike even its admirers as a joke when they're in darker moods. In such a view, science appears to be merely a series of prosthetics, donned by hairless apes. Telescopes are just farther-seeing eyes, radios merely a way of shouting loudly across mountaintops, microscopes simply large magnifying glasses, computers just a way to externalize and formalize our language—even impressive biomolecular armament like CRISPR are merely an extremely tiny and precise set of scissors. And the useful power of these prosthetics often fools us into overestimating our own powers, as if enhanced sight or hearing meant actual true understanding. Scientific research into consciousness is a perfect example: we both need a theory of consciousness in order to render neuroscience post-paradigmatic and at the same time seem stymied at every turn when we try to develop one.

Perhaps the irreconcilability of the intrinsic and extrinsic implies that the universe is not a closed system reducible to a model in the way we want it to be. In a sense, we scientists naturally gravitate toward thinking of this world as its zombie world twin—as merely a model, the manifestation of simple rules of physics and nothing else. We want to reduce everything to the extrinsic perspective, to a world that may as well be a complicated cellular automaton, a set of billiards. To be even more precise, we want the world to be a Turing machine, and it was Turing who said this best: "Logical computing machines can do anything described as . . . purely mechanical."[1] The person pursuing this idea most directly, and with the most success, is Stephen Wolfram, the physicist and computer scientist. Wolfram is currently using supercomputers to search through many simple rules for one that creates abstract structures that "look like" our physics.[2] If he ever found a simple rule that, if left to run by itself, would create a physics much like our world, then he would likely have found that elusive Theory of Everything. This is the ultimate extrinsic perspective—that reality is merely an abstract network of relations in which we observe regularities and call them the "laws" of physics.

And this vision of the universe has been immensely successful, scientifically—Wolfram hasn't found his theory, but beyond Wolfram, this hypothesis in one form or another has been championed since the beginning of science. Perhaps its greatest success is simply giving scientists confidence that things are understandable, that there are theories and truths out there waiting to be discovered, and that these truths are simple and elegant. But can this extrinsic perspective of the world ever truly be reconciled with the intrinsic? Or does it necessarily engender paradox?

An analogy is in order here. And I must point out that analogies can be dangerous, if one takes them literally—but they can also be usefully suggestive. The analogy involves science's queen, mathematics, which has been on epistemically shaky grounds ever since Bertrand Russell wondered: "Does the set of all sets that don't contain themselves contain itself?" If the set does contain itself, it shouldn't contain itself. If the set doesn't contain itself, it should. It's undefined. And this was the beginning of the collapse of certainty that ended in Gödel's theorems showing that formal systems built on axioms were necessarily incomplete and, not only that, that they could not prove their own consistency. This did not destroy mathematics, but it did confine it, fundamentally. It showed that there are doors mathematics cannot open, that there are walls it cannot climb, that it is hemmed in by an invisible force that one cannot see but only deduce.

And here's the analogy: we should ask if science itself has similar limits to its knowledge. Not limitations due to complexity, or difficulty, but rather based on fundamental foundational constraints. Perhaps it is too much to think of science as if it were exactly like a formal system of proofs operating under the laws of empiricism, but surely it is not a stretch to view science, in its algorithmic workings at uncovering truths of the world, as *somewhat* similar to a formal system. While scientific theories are usually not stated formally enough to be used to demonstrate this, we can imagine that science could be said to be complete if all questions about scientific theories can be answered, or, equivalently, if all statements made in some sort of hypothetical "language of science" can be decided as true or false (versus some being undecidable).[3] Perhaps science too contains the strange loop of self-reference, fatal flaws, holes where knowledge should be. Let us call this idea "scientific incompleteness."

The analogy of scientific incompleteness to mathematical incompleteness contains some striking parallels. When we model the universe from the extrinsic perspective, as Wolfram and others want to do, it appears entirely self-consistent, everything discoverable, empirically provable. It appears to be the same as the zombie world, and it therefore has no perspectives, no streams of consciousness, anywhere in sight. This is Thomas Nagel's "view from nowhere." And yet such an omniscient viewpoint appears to be incomplete. It is missing information. For among all the zombies of the world, when viewed from the extrinsic perspective, which one am I? Which one are you? *Why* are you that particular one? What rule binds you to it?[4] The extrinsic perspective is like a great map, utterly perfect in all its details, but upon which there is no label "YOU ARE HERE." And that is true for everyone everywhere, all at once. Such a map seems necessarily incomplete. And perhaps this is necessarily so, if reality outstretches our attempts to model it.

For isn't someone conceiving of this "view from nowhere"? The conceiver has a unique perspective, but under the most extrinsic of perspectives there are no unique perspectives, indeed there are explicitly no perspectives at all. Another example: In your average textbook that discusses the scientific method the experimenter makes an appearance at some point as part of the algorithm of testing and creating hypotheses. But experimenters in science are always thought of as being outside of the system they are observing and intervening on. Even just talk of "controlling variables" admits to this. More broadly, normally the observer and observed are also thought of as separate. One can "nest" this perspective in another extrinsic perspective that includes the observer, thus obliterating the distinction between the two as the observer is merely now a part of the same system. But this requires specifying some further extrinsic perspective. Which can in turn be nested into another one, and so on, ad infinitum. That is, there is never any true resting spot wherein the totality of facts are under consideration, since this always involves specifying some perspective, and a perspective, at least once we imagine it, contains an observer.

When experimenters are part of the experiment, does that not take us dangerously close to paradoxical waters? Wouldn't we expect to find difficulties in science *precisely* around observers, if the paradoxes that self-reference engenders in mathematics are kept in mind? For we seem to be having quite the difficulty in establishing a scientific theory of consciousness, which is exactly what we would expect if scientific incompleteness were the case.

For example, if we think of the independent experimenter as axiomatic to science, as one of its necessary postulates, it should be noted you can't pull yourself up by your own bootstraps and prove an axiom from within the system itself. In which case experimenters are irreducible. And there are indeed some arguments that observers cannot appropriately measure and distinguish all the states of a system they are themselves in, no matter if the system is classical or quantum.[5]

A metaphor. Imagine a perfect map of an island. And I do mean perfect—even though it need not be as large as the island, it is exactly to scale, such that every rock, tree, and even grain of sand is represented on the map, in incredibly fine detail. Astounding, but still, at first, conceivable. Now imagine that the map is on the island itself. What happens? The observer is now in the observed. For if we think on that perfectly detailed map, we see that it must contain, within it, a map of the map. And that map must also be perfectly detailed, and contain a further map of the map.[6] An infinite recursion. And what is a brain if not a map of the world? Like maps, brains represent the world around them, creating a world model. But the brain is a part of the world. And, again, to make the analogy between science and mathematics, this looks at least somewhat similar to Gödel's original proof: in it, the technique Gödel used to trigger paradoxes was a mapping which encoded statements about an axiomatic system into the system itself via something called "Gödel numbering." The encoded sentence (sometimes called the "Gödel sentence") that breaks these systems is always something like "This sentence is not provable" (Turing's version of this involved the question of whether a program would halt or not). Perhaps, similar to Gödel sentences, the brain trying to map itself triggers a recursion and concomitant paradox. One possible hint of this is that consciousness doesn't seem to lend itself at all to mechanical descriptions, which means that, when we attempt to embed experiences in the brain's map of the world, they seem to lose their essential and defining intrinsic properties, leaving us with only extrinsic ones, like neurons firing.

Interestingly, this highly skeptical view does not contradict any of the popular scientific theories of consciousness. For no existing theory of consciousness directly explains the property of subjectivity. Instead they focus in on the extrinsic properties of consciousness. This is true even of phenomenological theories. In IIT such properties are that it is structurally integrated, that it is composed, etc. All such properties are within the purview of contemporary science. Indeed, this view that science is incomplete may mean that we have to accept a weaker version of a theory of consciousness, one that, like IIT, may explain much about the structure of our phenomenology but little at all about the *why* behind phenomenology.

All this is a fantastic claim, to be sure. We are not in established science; we are closer to metaphysical speculation. Yet at the same time, there is actual suggestive evidence for this view, other than the usual endless philosophical difficulties that crop up around consciousness. For when we expand our scope broadly enough, we do see other examples of scientific incompleteness beyond consciousness.

First, there is the obvious paradox of "Why is there something rather than nothing?" While it comes with the connotations of late-night dorm-room debates, the simple truth is that the best scientists in the world cannot answer this question well—the physicist Lawrence Krauss wrote an entire book called *A Universe from Nothing*, and yet what he really put forward was a speculative account of how the universe of matter could arise from "merely" an unstable vacuum and the laws of physics.[7] But why is there a vacuum with these particular properties, and why these particular laws? Why any laws at all? It seems irresolvable, and for precisely the reason that Aristotle originally pointed out millennia ago: the paradox of the unmoved mover. An uncaused cause. A hole in the world, or at least, our explanation of the world, that has not truly been resolved to this day.[8]

As John Horgan argued in *The End of Science*, scientific theories as a whole are unfinished as they stand, with numerous gaps of knowledge still remaining, waiting to be filled.[9] However, are these simply incomplete in practice, and soon could be filled by scientific advancement? Or is some subset incomplete in principle? There is some reason to think the latter. After all, physical systems necessarily "inherit" problematic paradoxes from mathematics when one externalizes a Turing machine (normally a mathematical hypothetical) as an actual physical thing. Does it not arrive bearing all the same paradoxes? Although if true, does this really affect our understanding of the universe? After all, classically Turing machines are given an infinite amount of time and memory to complete their computational tasks—being physically manifested means this can't be true, so classical statements like "This computer program will halt" or "This computer program will not halt" are based on untrue assumptions, since, eventually, everything in this universe halts. However, there are even more radical views, including those of Stephen Hawking. Many have heard of or read Hawking's famously triumphant *A Brief History of Time*,[10] but fewer have heard of Hawking's later views, when he came to believe that scientific incompleteness is necessitated by the simple fact that science itself uses mathematics, and mathematics is itself paradoxical in nature.[11] He too was brought to this conclusion by a failure of science, although for him that failure was the inability to identify a Theory of Everything.

But beyond merely inheriting paradoxes from mathematics, researchers have found actual examples of undecidable properties in nature.[12] However, it is unclear if they exist in any problems that physicists are known to directly care about, like those that are empirically relevant.[13] Indeed, undecidability actually crops up in a lot of places. It has been proven, for example, that there is no best and decidable winning strategy for the complicated card game *Magic: The Gathering*, because finding such a strategy is at least as hard as solving the Halting problem.[14]

In 2015 a *Nature* paper caught my eye that I've never seen much publicized or discussed. It is as if no one knows what to do with it. What the researchers found was that a critical property in physics, which is called the spectral gap, is formally undecidable.[15] Specifically, for a given material there is some gap that exists between the energy levels of its electrons. What is the minimum gap for a given material? This is the kind of scientific truth that should be discoverable and that matters empirically. But you simply can't know the answer to this question, as it's formally undecidable. Even here, the issue turns out to be self-reference. According to Tony Cubitt, the first author of the *Nature* paper:

It also makes the self-referential structure of the proof evident: essentially, the proof constructs a Hamiltonian which can answer questions about spectral gaps through its spectral properties, then asks this Hamiltonian about its own spectral gap.[16]

That is, we already have some evidence that science, or at least the microphysical picture of the world, is incomplete, since it does involve (usually quite contrived) cases of recursion. If we conceptualize science as a great block of material, it may be riddled with holes. And from within, holes can only be identified by where the material curves, rather than the hole itself. If this conception of science is valid, then we should expect, of all things, consciousness itself to be undecidable, since it is science looking back at itself, and we know that self-reference is the surest way toward paradox.

Scientific incompleteness would indicate the double-edged sword that was Galileo's separation of the intrinsic from the extrinsic: the bracketing aside of the self would allow science to proceed unhindered, however, this is no guarantee that when subjectivity is added back into our conception of the world, when we try to merge the extrinsic and the intrinsic, we will be left with something 100 percent coherent.

We should overview some hypotheses related to scientific incompleteness. The first that stands out is in Douglas Hofstadter's books *Gödel, Escher, Bach* and *I Am a Strange Loop*, both classics of cognitive science as well as popular nonfiction writing. Hofstadter thinks that consciousness originates from the recursion of symbols within the brain itself (particularly of the most abstract symbol of the brain, that of "I," or the brain's symbol for itself).[17,18] This may or may not be true—although I do not see what is so special about symbolic recursion that it should *trigger* conscious experience, that is, necessitate it by its very existence. Nor exactly why animals without recursive symbol manipulation in their brain would be unconscious, nor do I think that if an NPC in a computer game were given a simple recursive self-symbol, it would spring into consciousness. However, regardless of the veracity of this theory, Hofstadter's view of how consciousness comes to be is tantalizingly related to, but is not directly required for, the view I am outlining, which instead says that science itself is incomplete and that consciousness is an example, likely *the central* example of this.

Similarly, there is the argument in Roger Penrose's *The Emperor's New Mind*, another classic book, that the human mind is non-algorithmic.[19] This is a decades-old argument—Penrose gives by far the best and clearest version of it,[20] but it goes back to the 1960s, originally proposed by philosopher J. R. Lucas.[21] It's therefore sometimes referred to as the "Lucas-Penrose argument." Specifically, this view hinges on Gödel sentences, which are the result of Gödel numbering. A Gödel sentence claims about itself that "this sentence is not provable in S" wherein S is some formal system. It is basically a complexified version of the liar's paradox ("I am lying to you right now"). If the liar's claim is true, then it's false, and if it's false, then it's true. If S the formal system is consistent (things are either provably true or false in it, they can't be both), and the Gödel sentence were proven in S, then that would imply it cannot be proven in S, since the statement "this sentence is not provable in S" would be true! That is, S holding that sentence to be true leads to a contradiction. Because if it were provable in S, it wouldn't be! However, from the outside, we can see that the statement is indeed true (after all, it's definitely true that the statement is not provable in S; rather, it's undecidable and triggers a paradox). But S itself can't ever come to this conclusion. Therefore, S is incomplete.

The Lucas-Penrose argument is that some form of this proof means that the human mind is not a machine, since we can "see" the truth of the Gödel sentence, even if S cannot. Although arguing this is even more complicated than the above explanation, since the Gödel sentence in question must refer to the formal system that is us! So we must postulate some formal system S that mimics the powers of our own human mind, like a Turing machine running a brain emulation, and see that it cannot prove the truth of its Gödel sentence; however, the Lucas-Penrose claim is that for this hypothetical Gödel sentence, we ourselves would know that the sentence is true, but meanwhile the formal description cannot know this—and therefore, the formal system S (like a brain emulation on a computer) is not really a full mimic of our mind, no matter how perfectly copied. It is an incomplete simulacrum.

This is a deeply controversial opinion, and understandably so, since it claims to be a direct proof that the mind is irreducible to the solely extrinsic. Some have argued strongly that the proof faces serious challenges, particularly around the issue that an assumption to the proof is that humans are consistent in the way formal systems are. Penrose has used this argument to propose a theory of consciousness based on quantum mechanics,[22] which has, again, proved controversial—especially since there is no obvious reason why quantum-based explanations lend themselves to solving the Hard Problem over standard neural-based explanations.

Regardless of its veracity, the Lucas-Penrose argument is actually distinct from the issues of incompleteness pointed out here (although neither is it contradictory). It may be that minds are non-algorithmic, but in the view I am outlining the difficulties around a theory of consciousness are an example of the incompleteness of science as a discipline—and this might hold true across a number of circumstances. It might hold even if Penrose is wrong and the mind is indeed simply some particular Turing machine. For this doesn't ensure a lack of paradoxes. Indeed, Paul Benacerraf, who was a philosopher of mathematics at Princeton, provided a mathematical proof decades ago that if you are a Turing machine, you cannot figure out exactly which Turing machine you are![23] The self-knowledge of which program is actually you is epistemologically closed to you. So if we were Turing machines, would this not place a limitation on a theory of our own consciousness? Not directly of course, as a theory of consciousness is more than knowing your own Turing machine. But let's imagine you had a theory of consciousness. If you had such a theory, and you were a Turing machine, couldn't you use it to figure out which one you were? Wouldn't a theory of consciousness allow you to place the "YOU ARE HERE" marker on the extrinsic map of the world? But that is precisely what Benacerraf's proof rules out.

Let us pause here. The problem with relying too much on proofs and cognitive puzzles and examples of paradoxes is that they become open for fencing with regards to philosophical and mathematical assumptions, which can vitiate the arguments quickly. And this is even true for all the proofs based on Gödel's theorems that apply to broader philosophical or scientific problems.[24] At the same time, I think philosophers and scientists have been wrong to dismiss this work in its entirety; the troubling proofs are more like a collection of signposts saying SCIENTISTS LOOK OUT! TROUBLE AHEAD. And the set of them all together makes it seem quite likely to me science itself leaves facts about the universe on the table. Which, we can speculate, includes facts about what it is like to be someone—a kind of true but unprovable statement. A theory of consciousness is like a Gödel sentence written in the language of science.

This would certainly explain the confusion around consciousness. Consider just the zombie argument: much like a Gödel sentence, it is true from the outside, thus explaining its initial conceivability and attraction. But once we insert ourselves into the argument, it becomes paradoxical. A pattern that plays out over and over with the mind-body problem.

There are, of course, echoes of this view in other places, some decades old. Its intellectual lineage starts, at least in the modern age, with the little-known 1952 book *The Sensory Order* by the famed Nobel Prize–winning economist, classical liberal, and author Friedrich Hayek, an intellectual giant of the twentieth century. Although more about communicating Hayek's ideas around psychology, some parts of the book seem to imply scientific incompleteness, albeit using very different reasoning. Hayek proposes the inability of the intrinsic to be reduced to the extrinsic in *The Sensory Order* this way:

> Mind must remain forever a realm of its own which we can know only through directly experiencing it, but which we shall never be able fully to explain or to "reduce" to something else.[25]

Later, in another paper, Hayek would lay out his reasoning more clearly:

> Any apparatus for mechanical classification of objects will be able to sort out such objects only with regard to a number of properties which must be smaller than the relevant properties which it must itself possess; or, expressed differently, that such an apparatus for classifying according to mechanical complexity must always be of greater complexity than the object it classifies. If, as I believe it to be

the case, the mind can be interpreted as a classifying machine, this would imply that the mind can never classify (and therefore never explain) another mind of the same degree of complexity. It seems to me that if one follows up this idea it turns out to be a special case of the famous Gödel theorem about the impossibility of stating, within a formalized mathematical system, all the rules which determine that system.[26]

Hayek's argument focuses on the complexity of the brain, since he posits that a classifying system must be more complex than what it classifies. I actually think this is either an incorrect or, at least, a weaker version of the argument that science is necessarily incomplete. But it is an early hint at the argument.

Aspects of the idea of scientific incompleteness also cropped up in debates around "new mysterianism" that swept philosophy of mind in the early 1990s. Mysterianism postulates that human minds are "cognitively closed" to an understanding of consciousness. The leading proponent of new mysterianism was Colin McGinn, author of *The Mysterious Flame: Conscious Minds in a Material World*.[27] However, McGinn's reasoning of the *why* behind cognitive closure also differs significantly—many of his examples of cognitive closure involve things like how rats, dogs, and humans all have different levels of cognitive closure, based on their intelligence, and this includes their conclusions about consciousness:

Properties (or theories) may be accessible to some minds but not to others. What is closed to the mind of a rat may be open to the mind of a monkey, and what is open to us may be closed to the monkey. Representational power is not all or nothing. Minds are biological products like bodies, and like bodies they come in different shapes and sizes, more or less capacious, more or less suited to certain cognitive tasks. . . . We could be like five-year old children trying to understand Relativity Theory. . . . We constitutionally lack the concept-forming capacity to encompass all possible types of conscious state, and this obstructs our path to a general solution to the mind-body problem. Even if we could solve it for our own case, we could not solve it for bats and Martians.[28]

That is, McGinn thinks the reason for cognitive closure is that we generally understand things in terms of perceived spatial relationships, and consciousness is non-spatial. But there are a lot of non-spatial things in this world, like bits or representations or other abstract ideas, which we seem to grasp just fine; and indeed, we are finely tuned by evolution to understand other minds, so appear to have the conceptual apparatus to do so, at least in one respect.

In comparison, scientific incompleteness has nothing to do with the limits of intelligence, nor the conceptual or sensory apparatuses that a certain species is equipped with. It is about reality itself containing innately undecideable properties, often triggered by paradoxical self-recursion of knowledge. Scientific incompleteness would explain why, when we sketch out merely the wiring diagram and functions of all the neurons, we arrive at a model like that of the Boolean networks we've been examining. Which is really a model of a zombie brain, with no consciousness anywhere to be found. Leibniz's mill. We might find the extrinsic properties of consciousness within such a model, as IIT attempts to do, but we cannot find the intrinsic properties of consciousness. We cannot fully envelop one world in the other. No matter how many partitions we stack or representations we compute or information flows we track, we cannot find subjectivity itself, merely its extrinsic correlates. Being itself possesses it, but our models of Being do not.

At this point Princess Elisabeth might slyly raise her hand. She'd ask how the intrinsic and extrinsic can ever interact, if they are indeed such different "substances." My tentative reply to her across the gulf of time is that reality itself has no problem of interaction—it "solves" it implicitly just fine—it is only when we seek to write down our own consciousness as a mechanistic model of equations that scientific incompleteness makes interaction seem impossible or paradoxical.

Another objection might be that just because something is an undecidable property does not mean that it is necessarily conscious—one might imagine a world with some sort of uncomputable physics without it necessitating consciousness. However, the idea of scientific incompleteness I'm putting forth here merely claims that consciousness itself is a *specific* undecidable property, one that is not captured when writing down the wiring diagram of the brain or even creating a simulacrum of the brain, not that undecidability automatically makes something conscious.

Ideally, of course, there would be firm evidence of this hypothesis, rather than merely support from a handful of paradoxes, in addition to some suggestive evidence that undecidable physical properties have indeed already been identified. So it remains merely a hypothesis awaiting deeper proof or disproof, with arguments for and against it.

Make no mistake: a clear disproof may come one day via a theory of consciousness that explains how the intrinsic *necessarily* arises from the extrinsic machinations of the world. Perhaps one day I will read such a paper, and the scales will fall from my eyes, and I will kick myself for not having thought of it. Or perhaps that is a young man's ambition, and any intellectual jealously will have long ago fallen away—indeed, is already falling away for me now, only in my thirties. I used to dream of that paper. In a way I will spend my whole life waiting for it. Perhaps its key insight will be communicable even in a single paragraph of text, and on understanding I will weep at its beauty. But if this does happen, it will imply a great revolution into unimaginable territory, territory I am sure no neuroscientist or philosopher walks on today.

So how seriously should we take scientific incompleteness? How should we relate to the universe if this is true? The essence of the human is our irrational hope, which springs eternal. Scientific incompleteness would be a slight crack in a door that refuses to close all the way, and it prompts us ever-hopeful humans to ask: Is the death of the physical brain actually the end of consciousness? For ourselves, for our lovers, for our sons and daughters?

Yet despite the refusal of the door to close all the way on hopes as old as humanity, scientific incompleteness would recommend no particular religion, no specific revelation other than uncertainty. It would mean that this world is like a great ancient aurochs, its breath steaming in the cold, its eyes a mirror of dark glass, its face bovine and unreadable. Decipher its expression if you can.

CHAPTER 10

How Science Got Its Scale

And yet, the book continues. Consider what follows to be a denouement of the possible, rather than the impossible, as we bracket aside consciousness for now. For even if we cannot reconcile the intrinsic and extrinsic perspectives on the world to our ultimate satisfaction, we can still make progress on evolving and shaping those perspectives. And indeed, recent work by myself and others has given us a much more formal and rich definition of what taking the extrinsic perspective even means, in the form of a theory of emergence.

This new definition of emergence allows us to answer not just long-standing scientific questions across almost every field, but also long-standing philosophical questions. Most importantly, I think it tells us something radical: that there is scientific justification for free will.

A word of warning. This chapter will involve some detailed and technical explanations around causation, although they should be understandable by everyone, no matter your background, and no more complicated than what we've covered previously. But we must do this; we cannot just jump to the end—a scientific justification of free will must be earned by deviling out the details of abstract concepts like causation, emergence, and information.

First, to understand what I mean about evolving the extrinsic perspective, let us zoom out and imagine the disciplines of science, all its subfields, and their relationships, taking the form of a great network—or perhaps a great tree, wherein each branch spans a different spatiotemporal level. Microphysics is at the bottom, then, moving up the tree, physics, and then chemistry and its subfields across the lower, thicker branches, followed by biochemistry, and so on. Perhaps such a tree fancifully recalls Yggdrasil, the Norse world tree, as it is the great structure that defines what things exist in reality. It defines what we intervene on, and observe, what entities we allow into the language of science.

Of course, no one knows precisely what Yggdrasil looks like. Only its shadow can be analyzed, via attempts to map science with various citation patterns of scientific papers.[1] Even more simply, we can view the tree of science as a ladder, where some fields are "higher up" than others:

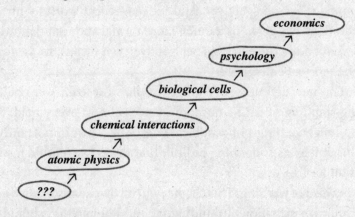

In this ascending ladder of abstraction, each step up moves us to another science (keep in mind that the ladder is fractal; if we zoomed in we'd see many more subfields of science at each step). Going up steps in this ladder means moving to a "macroscale," whereas we move down to some "microscale" below it. We might move all the way down to the ultimate microscale, that of microphysics, governed by atomic physics or even the wave function or who ultimately knows, or all the way up to the macroscale of markets and nations.

What exactly is a "macroscale," technically? Macroscales are things like cells or chemical bonds or animals or electrical circuits. They are dimension reductions of some underlying microscale state, dynamics, or elements. A dimension reduction is when some data is left out of a description, like a coarse-grain or average or grouping—e.g., the average height of students in a classroom is a dimension reduction ("dimension" here refers to the number of parameters needed to describe the system). There are near endless examples of macroscales; perhaps the most canonical is temperature, which is the average of the kinetic energy of particles in a system. Temperature shows how much can be left out of a dimension reduction: for any given temperature of a room, there are an astronomical number of particle configurations that could make it up, and we don't know which one of those configurations the air in the room is in.

Macroscales show up all over science; indeed, almost all the things we think of as normal science exist at some macroscale. For a neuroscientist, neurons are conceptualized as black boxes that take and sum inputs and then make a decision to fire or not, and what matters is often not what's going on inside, so the internal dimensions are left out of many neuroscientific models or experiments. All the current progress in AI is due to abstractly modeling neurons as such dimension-reduced black boxes. Real neurons are squishy quark clouds. But when neuroscientists analyze neurons, they are not conceptualizing them as squishy quark clouds, they are conceptualizing them as elements with some particular spatial scale (in the microns) and temporal scale (milliseconds). Neurons are often referred to as having "resting

states" or being "firing" or "active," all of which are dimension reductions of the underlying physics. In this manner, all scientific
fields occupy a certain range in space and time, a spatiotemporal range often little remarked upon, which defines what sorts
of elements and states that field includes in its analysis, as well
as what's in the models scientists in that field build of the world.

This extends to how we think about individual systems or objects in general. When taking the extrinsic perspective on something like a laptop, it could be described at the level of its quarks,
yes, but also at the level of its machine code, its compiling code,
and even the very high-level macroscale of the user-interface of
desktops and folders that we regularly interact with. The same
is true for your own body, which has multiple viable scales of
description, all the way from an astronomically complex quark
cloud, to cells, to organs and pathways, to your behavior as an
agent in the world along with your psychological states.

Certain scales of description appear more natural to us for
describing how individual systems operate. For example, cells
have a boundary (a cell barrier) that separates them from each
other, and they function as a unit—and following this, the scale
of cells is a natural description that biologists use all the time to
model and understand biological systems. Natural scales seem to
appropriately carve the system at its joints. However, there are
many other scales of description that we normally don't model—
it would make no sense to randomly assign cells to "supercells"
and view these "supercells" as fundamental units of biology. This
would be an unwieldly and uninformative macroscale. No biol-
ogist would want to work with such a model. Similarly, when it
comes to a computer, the levels of machine code and circuitry are
both natural descriptions. Yet it's easy to construct a terribly unnatural macroscale description of a computer, such as by measuring the average of sets of unrelated circuits. These macroscales

are so unnatural and useless we don't think about them, although they actually make up the vast majority of possible descriptions of any system. Despite it being a universal background of science, until recently there has been only minimal investigation into what separates natural scales of description from unnatural ones. This has led to confusion in sciences over what scales to understand their objects of study at.

Neuroscience itself has been racked with questions of scale ever since the doctor Camillo Golgi stained brain slices with silver, and then Santiago Ramón y Cajal used this technique to argue that neurons are separate units with gaps between them. Since then, neuroscience has been defined by what's called the "Neuron Doctrine," which holds that neurons are the functional unit of the brain's operations. Later, famous neuroscientists like Hubel and Wiesel[2] would further refine the neuron-centric view of the brain, arguing that individual neurons are selective for different features of the world (some pay attention to shapes, others colors, etc.). To this day, the individual neuron is almost universally thought of as the best scale of analysis for brain activity.

However, some neuroscientists have questioned whether individual neurons are actually the right scale to think about and model brain activity at. Of course, no one denies that the brain is made of neurons and glial cells (support cells), but rather many believe that some other scale is a more natural one for the description of the brain's actual function. Some scientists have proposed that the fundamental processing unit of the brain is minicolumns, which are composed of fewer than one hundred neurons and span the cortical layers. The idea is that these minicolumns are really the basic unit of brain physiology, and we are misrepresenting the brain by analyzing it at the level of neurons.[3,4] Not only that, there are plenty of levels above minicolumns at which one could analyze brain activity—and debates on what scales in neuroscience are important are long-standing and unresolved.[5] Indeed, trying to understand the brain at a macroscale is precisely what most neuroimaging does, focusing on things like local field potentials or even entire brain regions. Scientists failing to describe the brain at its most natural scale surely would explain a lot—it would mean that we neuroscientists have been observing and intervening on the brain in ways that aren't very meaningful to behavior or even perception.

All these issues are based on the fact that while science is the taking of the extrinsic perspective, this leaves open which extrinsic perspective to take. After all, there are a huge number of possible descriptions for any given physical system—some better, some worse, some more informative, or interesting—and we know this intuitively, but it's unclear what it means formally or mathematically. The existence of the many sciences means there is a reason for us to intervene in and understand the world at the level of the macroscopic rather than merely the microscopic.[6,7] Like asking how the leopard got its spots, we should ask: How did science get its scales?

Macroscales "Supervene" on Microscales

It is ironic that, despite macroscales comprising the bulk of all practiced science, science is generally thought of to be fundamentally based in reduction. Moving explanations down the ladder is what science supposedly aims at. Scientific reductionism has had a long history, and there is no denying that it has been enormously successful.[8] Indeed, so central are reductionist approaches to science that it is second nature for scientists to try to reduce a system to its smallest parts with the simplest behavior possible.

And I want to stress—I do not throw out some charge of scienticism or foolishness at the reductionist stance of most scientists. The idea that science requires or entails universal reduction is by no means obviously wrong. However, universal reductionism translates into an incredibly strong assumption: that scientists only use macroscales to understand the world because it is unwieldly (to the point of impossible) to talk about most of the things we care about down in a microphysical level of description.

Information theory gives us a useful way to talk about such reductionism: compression. When data is dimensionally reduced, it is compressed, and such a compression can be either lossless (like zipping a file on a computer) or lossy (information is lost in the compression). Most macroscales of science, like temperature, which loses the energy of all the individual particles, are lossy. The "null hypothesis" of how science got its scales is that science's chosen scales are simply a function of convenient compressions. Meaning that macroscales, pretty much all the objects and things we are familiar with, are useful merely because they are compressions.

One might imagine a race of alien superintelligences, committed reductionists, who could somehow model the world at a fine-grained scale of microphysics. According to universal reductionism, such aliens wouldn't miss out on anything, without caring about any macroscales, and could easily talk about and refer only to microphysical states.

The issue with the macroscales of science being merely compressions is that it seems that scientists should want to *avoid compression as much as possible*. After all, it's throwing out information about whatever they're studying. This seems to imply that, if they could be, scientists should be like these reductive aliens, and are simply inconvenienced into not being them.

But how do we know if macroscales really don't add anything beyond the convenience of compression? One of the clearest and most powerful arguments for this actually comes from a contemporary philosopher, Jeagwon Kim, who in 1998 wrote a book called *Mind in the Physical World*.[9] His argument is based around the idea of an abstract relationship called supervenience.

First, what is supervenience? When A supervenes on B what this means is that if everything about B is fixed, then everything about A is fixed. An example: the meaning of words supervene on their letters, a painting supervenes on its brushstrokes of colors, and Superman supervenes on Clark Kent. Another way of thinking about it is that there cannot be a difference in B without some difference in A, automatically. So, if Clark Kent dyes his hair, Superman has also dyed his hair, just automatically, since they are the same person. If a painter makes a change to their painting, they necessarily had to make a change to the brushstrokes. If you change around the letters of a word like "dictionary" to be "indicatory" you've also changed the meaning of the word, which supervenes on the letters.

This might seem an obscure logical entailment, but it happens to also describe the relationship between macroscales and microscales. Macroscales *supervene* on microscales. Another way of saying this: there cannot be a macroscale difference (like the temperature changing) without a microscale changing (something changes about the particles). The higher rungs in the ladder of science supervene on the lower rungs, implying that if we fix the properties of the lower rungs, the properties of the higher rungs will be fixed as well. And this spans all the scales of science: your psychological state presumably supervenes on your brain state, which supervenes on the molecular machinery of your cells, which in turn supervenes on your atoms. The ladder of science is actually a ladder of supervenience.

Kim's argument is that this supervenience relationship is philosophically problematic—not by itself, but specifically because of what it means for causation. He points out that we can always look at the microscale (like quarks bumping into each other), and certain things will seem to cause other things, but this can all be described down at the microscale, and the supervening macroscales are merely carried along by the causation going on in the basement of the world. That is, the underlying microscale causation *excludes* macroscale causation by rendering it unnecessary, merely a lossy (or at best, lossless) redescription—macroscales are totally redundant and can't add anything relevant.

Some have responded to this so-called "exclusion argument" by maintaining it simply cannot be accurate, since it is more like a *reductio ad absurdum*. Does causal influence really drain away as we go down the spatiotemporal hierarchy, spiraling down to microphysics and then whatever is underneath microphysics?[10] Are none of the macroscopic things you are familiar with, from cells to the people you love, anything but mere causal shadows? Indeed, Kim himself expressed ambivalence about this, hoping that it would only impact certain properties, like minds (I'm not sure why this is much better)—but he never was able to specify a good reason why the exclusion argument shouldn't apply to all macroscales everywhere, and therefore Kim's work represents one of the strongest arguments for universal reductionism.[11]

One fly in the ointment of the exclusion argument is that the supervenience relationship possesses an innate asymmetry—while there can never be changes in the macroscale without there being changes in the microscale, there can be changes in the microscale without changes in the macroscale. This is a property called "multiple realizability." Multiple realizability is why macroscales persist for so much longer than microscales. Much like the ship of Theseus, our body replaces the atoms inside of it, which flow through over a multi-year course in slow motion—for in a way the human body is both a solid and a slow-moving liquid, much like glass. The same is true even for non-living physical systems, things like hurricanes, or, indeed, ships that have had their parts replaced.

What does the multiple realizability of macroscales look like in the language of causal models? Macroscales in such models are simple: they are just a dimension reduction over some set of elements. Or, to put it plainly, you're basically just smushing two things (or more) into one thing, and using that in your model instead. Consider a model like the one below, with elements A and B (which have some inputs from some other parts of the system not shown, and outputs as well).

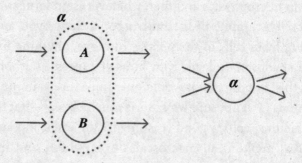

Both the causal models above are of the same system. But on the right we've replaced A and B with a new element, α, which represents the joint state of A and B. In order to do this, we have to specify precisely how α represents A and B—e.g., let's say α is a binary element (either 0 or 1, just like A and B), but it's in state 0 (α = 0) only if {AB} is in {00, 01, 10} and in state 1 (α = 1) only if {AB} = 11. That is, at a macroscale, we would read {AB} = 01 as α = 0.

This is an example of how macroscales are dimension reductions—they throw out information; here, what is lost at the macroscale is the difference between states like {AB} = 01 and

{AB} = 00. At the macroscale both are just $\alpha = 0$. It's also an example of multiple realizability. If at a macroscale the system is in $\alpha = 0$, this could mean A and B are in any of three possible states. They could be any of {00, 01, 10}—just as how, given a temperature in a room, the particles could be in many possible configurations.

A couple things to note. First, it's worth remembering that nothing is actually changing about the system at a macroscale; the only thing that's changing is the scale at which you're describing it at. Also note that there are all sorts of possible macroscales for a given causal model. We could have dimension-reduced {AB} in a totally different way. And also, one can do this to very complex models, like those representing networks, brains, or computer chips, and there can be many elements at the macroscale as different parts of the model are dimension-reduced, like below.

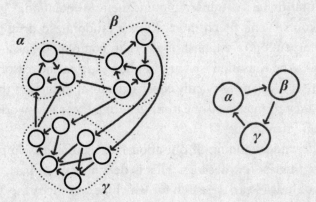

The existence of multiple realizability doesn't refute the exclusion argument by itself, but it does imply that there is a property that macroscales possess that by definition their microscales don't: their realizability by many possible microscales. In other words, there are many ways to be the same temperature, many ways to be the same "brain state," and many ways to calculate a particular input/output function via different algorithms. It is by definition impossible for a microscale to be multiply realizable—every state or configuration or dynamics down at a microscale is singular and unique. And this should give us pause, indicating, perhaps, an opening for macroscales to matter in a manner beyond mere compression.

The Growing Scientific Understanding of Causation

While causation is often thought of as a philosophical subject, it lies at the heart of science. Asking "What does what?" is the bread and butter of scientific research. In biology, this is expressed via a host of different kinds of interventions designed to understand what does what, like the up-or-down regulation of genes,[12] the optogenetic stimulation of neurons,[13] the perturbing of neural tissue via transcranial magnetic stimulation,[14] genetic knockouts,[15] and much more. Even a randomized drug trial is an intervention to establish if a certain compound has a causal effect.[16] And causation isn't just implicit to scientific reasoning—scientists regularly explicitly construct causal models of the systems they're interested in, from gene regulatory networks[17] to protein interactomes.[18]

The understanding of causation has significantly improved in just the last few decades. This is due, at least in part, to the work of Judea Pearl, research for which he was given the Turing Award (the Nobel Prize of computer science). Pearl formalized the idea of an intervention as key to analyzing causation.[19] An intervention is when an experimenter manipulates a variable in a system. Just as you might do an intervention to move a light switch up or down in your house, so might a scientist do an intervention to up- or down-regulate a gene. Pearl's work shows how intervention is qualitatively different than just observing. To understand causation, we cannot simply observe the correlation between things; we must specify some intervention that changes the system from the outside.

If you were to reverse-engineer some complex system, like an engine, or even a simple one, like when you figure out what light switches do what in a new house, you'd use interventions naturally—you'd tweak variables while holding others constant in order to figure out what causes what. Pearl formalized this approach to interventions mathematically as the application of a "do operator." Let's say you want to understand what the light switch in your new house causes. You would do(light switch = up), which would be flipping the light switch, which would, in turn, presumably turn on the connected lightbulb. This is sometimes called an "intervention-based account" of causation, or also sometimes a "difference-making account," since flipping the light switch made a difference to the state of the lightbulb.

Visually, we can think of such an intervention as performing a surgery on a causal model, where we fix the state of a variable and then see what happens. Generally, we start with a causal model in a particular state. Let's use one we're already familiar with:

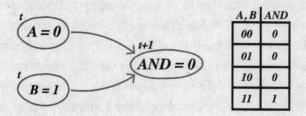

In order to know if A is a cause of the AND we can (hypothetically) "flip" the state of A to see what happens, which can be stated as do(A = 1).

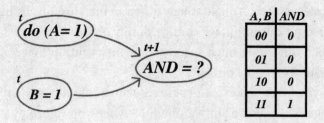

A, B	AND
00	0
01	0
10	0
11	1

Instead of AND = 0, we'd see an effect of AND = 1 following this intervention, meaning that it makes a difference what A's state originally was (and indicating there is causal influence from A onto the AND).

Assuming you have the correct causal model of a given system (leaving aside how one arrives at that, which is a matter of experimentation and inference), and that you can calculate the effect of interventions on it as above, this represents, according to Judea Pearl, "deep understanding." He writes:

> Deep understanding means knowing not merely how things behaved yesterday but also how things will behave under new hypothetical circumstances. Interestingly, when we have such understanding we feel "in control" even if we have no practical way to control celestial motion, and still the theory of gravitation gives us a feeling of understanding and control, because it provides a blueprint for hypothetical control. . . . We now see how this model of interventions leads to a formal definition of causation: "Y is a cause of Z if we can change Z by manipulating Y, namely, if after surgically removing the equation for Y, the solution for Z will depend on the new value we substitute for Y."[20]

That is, if manipulating Y entails a change in Z, this relationship means that the Y has some degree of causal influence on Z—just as shown above with A causing the AND's future state.

Notice that interventions capture something that mere correlations do not. Two different light switches might be highly correlated (a porch light and a lawn light), but you can tell that they are independent causal relationships when you start intervening on one and then the other.

Science has implicitly used this notion of causation ever since R. A. Fisher proposed the idea of a randomized experiment. For what does randomization do? In Pearl's terms, it performs a surgery on an equation in a presumptive causal model. For example, in looking at cases of heredity, the question of to what degree are outcomes of people nurture or nature is clouded by the joint and overlapping effects of both. That's why things like tracking twins who were separated at birth is so important in biology.

When I refer to causation, this is to what I refer: the influence of one variable in a model on another, which we can separate out from correlation by performing interventions. And as Pearl points out, once we have a causal model, we can conceptualize not just actual performable interventions, but hypothetical or imaginable ones—that is, we can reason counterfactually.[21]

Some philosophers think that causation is an incredibly deep metaphysical mystery. And perhaps its ultimate nature is, and we can only be content with a contextual understanding of causation depending on the type of causal model we're examining.[22] However, the new mathematics of causation obviates metaphysical questions by simply sidestepping them. Causation, at least in systems amenable to representation as the sort of causal model we're discussing, simply *is* the dependencies between variables (once interventions separate out any confounding factors).

This more precise understanding of causation allows us to phrase Kim's exclusion argument very precisely. If universal reductionism is true, it should always be preferable to "fine-graining" a causal model (as in increasing its dimensionality by adding more details). Moving down the spatiotemporal ladder should always give a better understanding of what causes what, never a worse one.

This is a fact amenable to investigation by examining measures of causation and how they behave as we describe them at different scales. And this is precisely the approach I and my co-authors introduced in 2013 in the form of a theory that tells us when to reduce, and when not to, in a manner that can end the questions of scale that hound and bedevil scientists.

Causal Emergence

During my PhD work on Integrated Information Theory, it was our job, myself as a graduate student along with a couple of post-doctoral researchers, to shore up the theory, to give it as much argumentative and intellectual firepower as we could. So various people worked on different parts of it, since the theory, as you know by now, is so complex.

One particular problem kept cropping up: we seem to experience things at a certain temporal scale (milliseconds to seconds), as well as some particular "spatial" scale phenomenologically (we cannot see things in infinite detail, etc.), and additionally neuroscience itself, when examining the presumable neural correlates of consciousness, chooses correlates at a spatiotemporal scale very far from fundamental microphysics. But doesn't the reductionist viewpoint preclude this? How does consciousness get its scale?

I wanted to solve this issue to help shore up IIT, but I also felt that it went far beyond IIT. If consciousness could really exist up at a higher spatiotemporal scale, why? Didn't that indicate a more general phenomenon? After all, according to the exclusion argument, all causal influence should drain away to the bottom microscale. And if it didn't, it hinted at a theory of emergence.

This is what I convinced my colleagues to work on: a general mathematical method to separate natural from unnatural macroscales by examining causation at different scales in systems. This research, which comprised the bulk of my PhD, eventually became what is now called the theory of causal emergence. And it tells us a surprising fact: macroscales can have causal influence above and beyond the microscales they supervene on.

How is this possible? The theory of causal emergence begins by acknowledging that there are degrees of causation. Things can have more or less causal influence. This is precisely how humans naturally talk about causation, as when we say, "The tree fell down in the storm, but mostly because it was hollowed out by termites before that." We're all familiar with how drugs might treat a disease, but not fully, or always. That is, some causal relationships (the dependencies between variables in a causal model under interventions) are stronger, while others weaker. There are a bunch of synonyms for this, from "causal influence" to "causal power" to "causal strength" to "causal control" to "causal informativeness" to "a better causal explanation"—take your pick.

Causal emergence occurs when macroscales have more causal influence than their underlying microscales over the exact same events. To actually measure causal emergence, we must pick a particular measure of causation in order to assess the degree to which macroscales improve causal influence. And when it comes to measures of causation, there have been many throughout history. Indeed, scientists, mathematicians, and philosophers have not all agreed upon one universal measure of causal influence, although plenty have been proposed. Instead, various overlapping definitions have been given, some of which capture different aspects of causation. This is much like the advances in complexity theory in the 1980s and '90s, wherein a number of different related measures of complexity were introduced. Even to this day, there is no ultimate agreement on what exactly "complexity" means mathematically—it really means a couple of things, different aspects of which are captured by a family of measures that often agree and occasionally disagree in interesting ways.[23] The exact same is true of causation. It is very possible that there will never be universal agreement over which measure of causation is best, since what we mean by "causation" sometimes differs in subtle ways, and there is no single equation that perfectly captures every single aspect of how we use the word—but this does not mean that there cannot be a science or mathematics of causation, just as there is with complexity.

And, luckily, causal emergence doesn't require finding the one true measure of causation. Examples of causal emergence can be found in even the simplest and oldest measures of causation—it is a general phenomenon concerning causal relationships, irrespective of exactly how you measure them.

Let's consider David Hume's definitions of causation from back in 1748, often recognized as laying the groundwork for almost all further thinking about causation. Hume gives two different definitions of causation, although the first is more famous, since Hume developed it substantially: "We may define a cause to be an object followed by another, and where all the objects, similar to the first, are followed by objects similar to the second."[24]

This is sometimes referred to as the "regularity account" of causation—if two things are paired with one another enough (say, burned toast and toaster ovens), then we eventually say that one is the cause of the other. There are all sorts of problems for this definition of causation, but what's interesting is that even in this very early definition the possibility of causal emergence appears—for are not macroscale relationships far more regular than their microscales? Indeed, the states of microscales often never reoccur, while the states of macroscales regularly do. And since macroscales are multiply realizable, effects can be paired better with their causes at a macroscale.

Let's examine a simple case of this in a causal model, explicitly using Hume's regularity account of causation. Since it will be boring and uninteresting if everything correlates perfectly, let's add in a bit of noise to the model. That is, the dependency of one element on another element will be defined in terms of probabilities. All this means is that we let the truth tables that govern the elements use any value in between 0 and 1 for calculating their outputs. Imagine a causal model wherein A inputs to {BC}. Since it is a probabilistic model, these inputs can lead to different outputs. For example, if the input A = 0, this will likely lead to the output of B = 0, but only 70 percent of the time; the remaining 30 percent of the time the output is B = 1 (and the same rules for C). The microscale of this system is on the left, the macroscale on the right below, along with the tables describing the binary inputs at each scale, with the outputs given in terms of the probability of being 1 (at the microscale, B and C share the same mechanism).

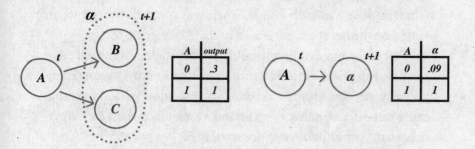

We can compare the two scales using Hume's regularity account by looking at the pairings between states. Imagine the pairings at the microscale we might observe: we might see A = 0, but since both B and C now only have an independent 70 percent chance of being 0, we might see A and B assume another state—the highest pairing from A = 0 is {AB} = 00, which is what A = 0 is most strongly associated with, but this occurs only 49 percent of the time. Meaning that, at the microscale, states don't regularly follow other states.

However, at the macroscale, there is almost perfect regularity. This is because α groups over the states of {BC} such that any of {00, 01, 10} are described instead as $\alpha = 0$. Therefore, $\alpha = 0$ is strongly associated with A = 0 (91 percent of the time, since $\alpha = 0$ anytime that B and C equal 00, 01, or 10, which they have a cumulative probability of 91 percent of doing so, given A = 0). According to Hume's original definition, this is a case of causal emergence, since the macroscale has greater regularity between its elements than the underlying microscale that it supervenes on.

Perhaps this is a weird by-product of the regularity account (which we might be skeptical of to begin with, given its lack of sophistication)? It turns out no. For there's another definition of causation that Hume gives "where, if the first object had not been, the second never had existed."[25] This definition by Hume is sometimes called the "counterfactual definition of causation." (Imagine living in such a time, when you could throw out what would be the heart of centuries of debate in a single line!)

One of the most influential philosophers of the twentieth century, David Lewis, writing more than two hundred years later, formally defines Hume's account in this way: something is a cause c of e if, given that events c and e took place, then if c hadn't occurred, e would not have occurred.[26,27]

Using Lewis's definition of causation, it is again easy to find a case of causal emergence. Here's another causal model set up a bit differently—this time without noise, but with redundancy instead. It's actually our old friend, the AND gate.

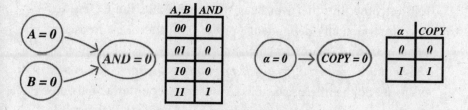

A, B	AND
00	0
01	0
10	0
11	1

α	COPY
0	0
1	1

We can again redescribe this system at a macroscale, wherein now the inputs {AB} are grouped into α, such that {00, 01, 10} are considered α = 0, while 11 is considered α = 1. This transforms the AND into a COPY gate that just copies the input it receives, so if α = 0 then COPY = 0 (although the element is not actually changed; this is just a different way of describing what's going on). Let us examine the Hume/Lewis definition of causation in the above case. To compare the two scales, we want to know if AND = 0 counterfactually depends on {AB} = 00, and if, at the macroscale, COPY = 0 counterfactually depends on α = 0. To assess this, we can make use of the "do" operator at both scales:

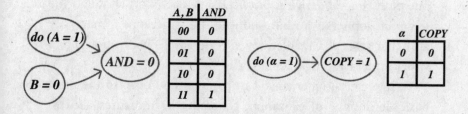

A, B	AND
00	0
01	0
10	0
11	1

α	COPY
0	0
1	1

We can see the results of these interventions and make judgments about the counterfactual definition of causation at the two scales. At the microscale, if not {AB} = 00 (e.g., if {AB} = 10), we can see that AND = 0, just as before. Meaning that AND's state being 0 is not counterfactually dependent on {AB} according to this definition of causation. However, if not α = 0, then necessarily α = 1, which in turn means COPY = 1, which is a different outcome. In other words, at the macroscale, the COPY = 0 is counterfactually dependent on α = 0, but at the microscale, AND = 0 is not counterfactually dependent on {AB} = 00, *even though both are just differing descriptions of the same occurrence of events.* So the microscale fails to satisfy the counterfactual definition of causation, and is not a cause at all, while the macroscale does satisfy the definition—a simple case of causal emergence, since the macroscale is a cause while the microscale is not.

These may seem incredibly abstract examples. Do they translate into the real world? To see, let's apply these ideas to the question of whether your behavior is caused by your neural states. This probably means that small changes (via a hypothetical intervention) to your neural state would have led to differences in outcome. However, small changes to your atomic state (via a hypothetical intervention) would not have led to changes in behavior. So in what sense were your atoms a cause at all? Some philosophers have pointed out similar scenarios to this, focusing specifically on the idea that mental states are necessary for outcomes but underlying physical states are not, although the few philosophers who have noticed this difference in necessity at higher scales have failed to generalize this observation into a wider understanding of emergence.[28-30]

While causal emergence can be identified even in the most basic definitions of causation, generally the measures used to determine its existence are more complex than Hume's original definitions, often making use of information theory.[31-35] Much of the work in the causal emergence literature has been done using what's called *effective information.* Effective information is a measure of how much difference the average intervention makes in a causal model. It is calculated by randomly intervening on elements and then measuring the information transmitted, using information theory.

I and my coauthors showed back in 2013 that effective information could increase significantly at macroscales under the right
circumstances.[36] The first system we showed this with was more
complicated than some of the causal models we've looked at, in
that it has four elements, but the basic principles are the same.

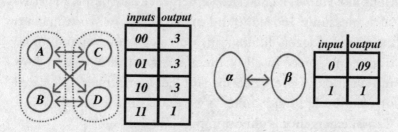

Notably, the effects of interventions in such a system are very
different at the macroscale compared to the microscale. Microscale inputs are both almost never necessary and also may
only unreliably and irregularly lead to particular outputs—for
example, if AB = 00, then do(AB = 01) makes no difference. At
the macroscale, however, interventions look very different: they
are almost always necessary for outputs, and they also trigger
them reliably, meaning that do(α = 0) is very likely to lead to β =
0, and β counterfactually depends on α. This is reflected by the
information transmitted by the average intervention, which is the
effective information (1.15 bits at the microscale, but 1.55 bits at
the macroscale).

We also later showed causal emergence using the integrated
information, the measure at the heart of IIT.[37] This answers the
original question of why it is (at least, according to IIT) that consciousness doesn't necessarily have to be associated solely with
the nearly instantaneous quark cloud of the brain.

If causal emergence only occurred due to some particular
quirk of a single measure, such as because of how the effective

information was calculated off of the average intervention, then it would be sensible to be skeptical of whether causal emergence is a real phenomenon.[38,39] But this is not the case. Rather, there are dozens of measures of causation, many of which are highly related (and many of which are modern updates of older notions like Hume's), and causal emergence is widespread across such measures; indeed, I and my coauthor Renzo Comolatti examined a dozen different measures of causation and found causal emergence in all of them, irrespective of exactly how interventions were performed or counterfactuals calculated or what background contexts were used in causal analysis.[40] Causal emergence is almost impossible to avoid.

This changes our conception of what emergence is. It's not some rare exotic scenario that occurs only at the margins of physics. Previously, examples of emergence relied on slogans like "more is different,"[41] which, while intriguing, has so far not lent itself to provable examples of emergence, acting more like a suggestive metaphor.[42]

Causal emergence is not as sexy as some magical physics-violating new properties popping into existence at the macroscale. That's a good thing. Causal emergence is not magic, just math. And it's so common it's nearly everywhere you look. The reason for this, it turns out, has to do with how your cellphone works.

Causal Emergence as Error Correction

Isaac Asimov's *Foundation Trilogy* imagines a fantastical science of history called "psychohistory," which the character Harry Seldon uses to predict the galactic empire's inevitable downfall. Asimov refers over and over to how Seldon's predictions only work at the aggregate level (a macroscale), because the individual actions of people cannot be predicted, due to their inherent randomness and chaos—although they all average out at the aggregate. That is, the aggregate performs what's called "error correction."

The technical definition of error correction comes from information theory. It underlies basically all communication technology, and it was originally discovered by Richard Hamming, who along with Claude Shannon was one of the founders of information theory (they even shared an office at Bell Labs). Even at a psychological level you yourself engage in error correction all the time: if I wrie whle droppin leters ther's a gd chanc ou'll stil undertand wat I'm saing.

The idea of error correction in information theory is relatively simple. It starts by formalizing communication as being something that occurs along a "noisy channel." Imagine a telephone line that goes in and out as you're talking—that's a noisy channel. But let's make it even simpler and imagine it's a communication channel over which you're just sending 0s or 1s, rather than entire words. The noise in the channel can be presented as some probability that the 0 you're trying to send "flips" to being a 1 for the receiver, or vice versa. This bit-flip probability could be quite high—it might mean that if you tried to send 0110 it would come out as 1101, or some other garbling, which is obviously problematic for communication.

What Hamming showed was that you can overcome this noise by encoding your message in a way that takes advantage of redundancy.[43] As a simple example, rather than just sending a 1, you could send 111 to mean 1, and 000 to mean 0. While you'll communicate slower, the receiver will have a much higher probability of understanding what you're trying to say, as the probability of flipping all three bits is much lower than the probability of flipping a single bit. And on the receiving end, the strategy is to group the outputs—treat {000, 010, 100, 001} as being a 0, and then {111, 110, 011, 101} as being a 1. You might sometimes still get the message wrong, but overall the errors in communication will be significantly reduced.

Notice, however, that this is already a lot like you're analyzing the message up at a macroscale! After all, you have your two "macro-messages" {0, 1} that can come out of a language of micro-messages {000, 001, 010, 011, 100, 101, 110, 111}, so there's a dimension reduction, and the macro-messages are multiply realizable, as the receiver could receive 0 from either 000 or 001 or 010 or 100.

Since information channels are merely some input/outputs with associated probabilities (much like the truth tables we've been discussing), there's no reason not to apply the same thinking to causal relationships, which are mathematically very similar to information channels—indeed, because the effective information is so close to central measures in traditional information theory, I was able to show directly that much of the same mathematics applied.[44]

What does it mean to have "errors" in causal relationships? Imagine your new house contained two light switches, each of which turned on a different light. One light switch is deterministic and infallible at turning on its connected light. It always works. This is an error-free causal relationship. But now imagine another light switch that is indeterministic (or "noisy"), meaning that the light only sometimes turns on when the switch is flipped.

In our work, we formalized the difference between the two as the amount of *determinism*, the degree to which a causal relationship is deterministic. Assessing the degree of determinism is like asking: If the light switch is on, what do we know about the bulb? For a deterministic light switch, we know for certain the state of the bulb just by looking at the switch. However, for an indeterministic light switch, we know little about the state of the bulb. It is as if the switch is trying to influence the bulb, but there are "errors" where the "signal" (the cause) goes missing or gets swapped.

This is just one of two types of errors in causal relationships. The other kind of error is called "degeneracy" (sometimes also called "redundancy"). The term "degeneracy" comes from biology, when many different genetic configurations lead to the same phenotype, or many different neural states lead to the same behavior—indeed, the degeneracy of the cortex is something Gerald Edelman himself wrote about. Degeneracy can be formalized by looking at how often different inputs lead to the same output.[45]

Imagine now a lightbulb controlled not by one, but by two switches, either one of which can be used to turn it on or off. Assuming the relationships are deterministic, where the "errors" are here may at first not be clear, but it makes sense if you think about it: due to the degeneracy, if the bulb is on, it may be because one of the switches is up, or it may be because both are up. All you know from the output (the state of the bulb) is that both light switches can't be off, which is less information than in the case where there is only one switch that uniquely controls the bulb, in which case you would definitely know the state of the switch from looking at the bulb. In other words, there can be errors in relating effects to particular causes, which is different than the errors of indeterminism, which are potential errors in relating causes to their effects.

Both determinism and degeneracy matter a great deal to causal influence. Pretty much all measures of causation are based in some combination or relation of the two.[46] In general, causal influence increases when determinism increases (effects reliably coming from causes), and decreases when degeneracy increases (as the causes become more redundant). Note that this maps well onto the traditional philosophical understanding of causation, such as how causes depend on sufficiency and necessity.

So then what is the causal equivalent of an error correcting code in information theory? A macroscale causal model that increases the determinism and/or decreases the degeneracy. The more error correction there is at a macroscale, the stronger the causal influence will be in comparison to the underlying microscale, and therefore the greater degree of causal emergence there will be.

If what Richard Hamming realized is that the secret to sending a message in the presence of uncertainty is redundancy, then causal emergence is the realization that creating strong causal relationships in the presence of uncertainty can be accomplished via multiple realizability—for the exact same mathematical reason. And microscales by definition cannot have this property of error correction, as everything is a one-off—nothing is multiply realizable down at a microscale.

Philosophers and scientists will sometimes distinguish between what they call "weak" and "strong" emergence, with the weak kind being uncontroversial but also unexciting, and the strong kind being so controversial that only a slim minority believe it exists.[47,48] Weak emergence just means that macroscales are difficult to predict or understand in practice, but are reducible to the microscale in theory. Strong emergence means that new irreducible properties are added up at the macroscale, which looks a lot like it requires new physical laws for just the macroscale. Yet these present us with bad choices, as weak emergence is relatively uninteresting, whereas strong emergence is too interesting!

Causal emergence offers a middle road, being neither weak nor strong emergence. It holds that the elements and states of macroscales are reducible to underlying microscales without loss, but that the causation of the macroscale is not. At the same time, this "extra" causation is not unexplainable or mysterious. It's just error correction.

It Works in Practice. But What About in Theory?

Given the implications on offer from a theory of emergence, it is little wonder there are some possible objections to the idea of causal emergence. In my experience, they require either (a) quibbling over a particular measure of causation, which is made extremely difficult since so many different measures of causation show the phenomenon to some degree or other, under many different contexts and assumptions,[49] or (b) purposefully obliviating the differences between a microscale model and a macroscale model, as in using macroscale probabilities or counterfactuals down at the microscale, or allowing the microscale to be a dimension reduction (i.e., not be a real microscale at all), all of which can be described as leaking information between macroscale and microscale models (which makes zero sense, since the whole point is to compare and contrast different scales).

However, one serious objection is actually a consequence of the theory itself. According to the theory, causal emergence is impossible if there is no uncertainty anywhere to be found, since there are no errors to correct. Which leads to the natural question: Where does uncertainty in our models of the world come from? Whither the errors?

This is essentially an empirical question. The universe could turn out to be a bunch of different ways, perhaps determinate, perhaps indeterminate (operating via probabilities). In practice, there is uncertainty almost everywhere in science. For example, in biology, randomness can come from how cell molecules exist in the presence of Brownian motion,[50] or the stochastic nature of ion channels.[51] And causal emergence only requires that probabilities (uncertainties) crop up *somewhere*. We can go back to Stephen Wolfram's attempt to find some simple rule which generates physics—the vast majority of the rules that Wolfram has explored, including some he thinks underlie our own physics, contain degeneracy (uncertainty about the past); indeed, he speculates that this degeneracy is what underlies various counterintuitive aspects of physics itself.[52]

Down at the microscale of the world our knowledge is incomplete: we still cannot reconcile quantum physics with general relativity, and gaps remain in physics itself, as string theory has not fulfilled its promise as a theory of everything.[53] Quantum physics is based on probability distributions, and notoriously contains indeterminacies around measurement, and only controversial interpretations of quantum physics, like so-called superdeterminism[54] or other hidden-variable theories, seem to get rid of this irreducible uncertainty. Proponents of these seem to be in the minority in physics. Additionally, there are some important proofs that show that uncertainty in physics may be necessary and unavoidable;[55,56] similarly, scientific incompleteness would necessitate uncertainty somewhere in, at minimum, our models of the world.

Furthermore, in any open system there is necessarily indeterminism. Even the rays of the sun might occasionally interfere with the workings of a computer, adding a tiny bit of infinitesimal noise. If a scientist goes to make a causal model of a computer, it seems ridiculous to stipulate that they must include the entire atomic state of the sun. The more natural thing is just to add in a small noise term somewhere in the model to represent the fact that it's an open system, or that components may break, and so on. That is, it is natural to treat causal relationships as local, and therefore noisy, rather than trying to incorporate the entire state of the universe in each model.

A committed reductionist might bite the bullet and maintain that all causal models should indeed include the entire state of the universe (while also assuming that physics has no other source of errors in causal relationships anywhere). Yet there is even a problem with this extreme view—it eliminates the notion of causation entirely. An analogy: entropy always increases in closed systems, but it doesn't always increase in open systems. Here on Earth entropy can be combated; indeed, that's exactly how life functions, like every time you clean up the kitchen or make your bed or eat food. But when taken as a whole, across the universe, entropy is always increasing. The ability for entropy to decrease only exists in open systems—that is, locally.

Similarly, when the entirety of the universe is taken as a whole, there is only exactly what happens, not what can or should have happened (no counterfactuals or interventions), and so causation itself disappears. According to Judea Pearl:

> If you wish to include the entire universe in the model, causality disappears because interventions disappear— the manipulator and manipulated lose their distinction. However, scientists rarely consider the entirety of the universe as an object of investigation. In most cases the scientist carves a piece from the universe and proclaims that piece *in*—mainly, the *focus* of investigation. The rest of the universe is then considered out or background and is summarized by what we call boundary conditions. This choice of ins and outs creates asymmetry in the way we look at things, and it is this asymmetry that permits us to talk about "outside intervention" and hence about causality and cause-effect directionality.[57]

So at this point, the committed reductionist must conclude that causation itself does not exist (meaning, unsurprisingly, that causal emergence wouldn't either). In other words, only the most extreme and specific views of ontology and epistemology—like assuming that there is no indeterminism or degeneracy anywhere in the universe, and, even if there were, insisting that no subset of the universe can be modeled independently in any serious way as an open system and therefore advocating that causality itself does not exist—can lead us away from the conclusion of causal emergence.

Natural Scales of Description

A theory of emergence should be *useful* to science beyond its metaphysical conclusions—that's one way we know it's true. So what can we use a theory of emergence for? This takes us back to the original motivation of this chapter: answering how science got its scale. More specifically, causal emergence allows us to pick out natural scales of descriptions of systems—that is, good spatiotemporal scales to make causal models at.

It's worth noting this can work both ways. A theory of emergence is also a theory of reduction. Simply reverse the sign! Causal reduction is the opposite of causal emergence—and indeed, the theory tells us when such reduction is appropriate. Because of this, finding the right scale to model a system at is very much like you are looking at the causal structure of the system at different scales, like focusing a camera—when you hit the natural scale of description, the system "snaps" into focus and what causes what becomes obvious.

Remember that scientists want to avoid compression as much as possible (that is, they want as many elements and variables in their models as they can have, so they avoid throwing out information). However, they also want causal models where the elements or variables show a high degree of causal influence over one another. The theory of causal emergence therefore offers an elegant definition of naturalness: scales where the macro performs the most error-correction for a minimal tradeoff in terms of dimension reduction. That is, a natural scale of description is the point where the most minimal of dimension reductions leads to the maximum gain in terms of the degree of causal influence parts of the model have on one another.

Finding such points can be done in a few ways. Some measures of causation already contain within them an implicit "size" term (as effective information does) that makes them sensitive to dimensionality reduction—integrated information is similar. In these cases, to find a natural scale of description the modeler can just maximize that measure (like finding the maximum of effective information or integrated information across scales). In other cases, where measures don't already combine the two, causal influence can be balanced against compression. Imagine shaving off the dimensions of a system (leaving more and more out, moving higher and higher in spatiotemporal scale, etc.) and seeing consistent gains in terms of the causal influences in your model, until eventually hitting a point of diminishing returns—this is the sign you've dimension-reduced enough and are close to a natural scale of description.

Note that only an astronomically slim minority are good dimension reductions that show causal emergence, and in practice we often need heuristics to identify causally emergent scales.[58] Using such heuristics, we've shown that causally emergent macroscales are more common in biological systems compared to technological systems.[59] This makes sense because biological systems have a lot of uncertainty, and so there are errors to correct. Indeed, across more than one thousand analyzed species, we've compiled suggestive evidence that causal emergence in protein interactomes (the bindings of various proteins in cells) increases as evolution goes along,[60] in that eukaryotes show a greater degree of causal emergence than prokaryotes. This might have something to do with the fact that evolution faces the challenge of constructing reliable causal relationships (such as between a chemical gradient and gene expression) in the face of the extreme noise and uncertainty of the physical world at the level of the cell. Imagine having to construct precise clocks out of Legos that phase in and out of reality, and that is essentially the problem biology faces.

Reconsidering the macroscales of science, like chemistry, biology, or economics, we can now see such descriptions of the world as what they truly are: error-correcting encodings of physics. That is, macroscales are not merely useful because we don't have the memory or computational resources or observation capabilities to understand everything at the microscale—in fact, we want to minimize compression as much as possible.

In a way, scientists search across the possible extrinsic perspectives on a system for the one that engenders the most deep understanding, particularly of what does what. At the same time, they are trying to avoid compressing too much and want models as large as possible. At natural scales of description, scientists' interventions are more certain and compression is minimal—a balancing act. So how science got its scales turns out to be an implicit search across the spatiotemporal hierarchy of the world, a search carried out by millions of scientists trying to find reliable causal influences, as if they were ants searching for rewarding sources of food and carving out tunnels on top of tunnels.

CHAPTER 11

The Scientific Case for Free Will

When William James was a young man, he was nearly bedridden for three years by a crippling existential depression. It was due, at least in part, to his inability to believe in free will, and therefore his questioning of his own will, his own purpose, his own meaning. His depression ended the day he came to believe that the question of free will was impossible to answer one way or another. Without proof either way, he realized the question was totally open and unanswerable, and therefore he could choose to believe in it, writing famously in his diary that "My first act of free will shall be to believe in free will."

James may be an extreme case, but belief in free will matters significantly for behavior, and for our culture. In a number of studies the belief in free will has been correlated with all sorts of positive psychological traits, like greater gratitude, more life satisfaction, less stress, reports of a more meaningful life, a greater urge to pursue meaningful goals, and even more commitment in relationships and a greater tendency to forgive.[1]

And what does it mean to not believe in free will? What is the impetus of this belief? I think, fundamentally, that it is simply taking the extrinsic perspective of the world as literally as possible. It is to look at the world and see only its mechanisms, its cogs and wheels, and see no place for the intrinsic perspective to matter at all. William James was suffering from a sickness of the extrinsic perspective.

Of course, plenty of thinkers have argued that free will does exist, often taking a "compatibilist" position wherein the concept of free will is compatible with the current laws of physics. However, there's a lot of contemporary skepticism for these positions, because as a strategy they often define free will in a way that is ultimately unsatisfying, such as by lopping off necessary or intuitive parts of it, and then showing that mangled definition to be true. Now, it is unavoidable that our naïve and untechnical definition of free will (the definition we start with) will change, so that's not the problem—the problem is that existing arguments for free will give up too much.[2]

However, the idea of causal emergence complexifies our extrinsic accounts of the world, and proves universal reductionism untrue. From here we can identify a definition of free will that's actually worth wanting. Why does causal emergence matter? Because while the exclusion argument is mostly confined to the philosophical literature, it is the academic teeth behind a lot of the more public-facing arguments against free will. If you've ever heard things like how we have no free will because we are merely trains running along the tracks of physics, such expressions are, at least in part, a statement of the conclusion of the exclusion argument, which is then simplified in its popular presentation.

If such views were true, it would indeed be a depressing state of affairs. The philosopher Jerry Fodor summed it up: "If it isn't literally true that my wanting is causally responsible for my reaching and my itching is causally responsible for my scratching, and my believing is causally responsible for my saying . . . If none of that is literally true, then practically everything I believe about anything is false and it's the end of the world."[3]

But if the exclusion argument is the academic teeth behind the more popular anti-free will arguments, we've already seen that causal emergence throws the conclusion of the exclusion argument into serious doubt. In our original introduction of the idea of causal emergence, which was based on identifying cases where the macroscale has greater effective information than the microscale, we maintained that the macroscale excluded the causation at the microscale.[4] That is, it flipped Kim's exclusion argument via intellectual judo: since we know that the macroscale is a better description of the causation governing two descriptions of the exact same occurrence, then what do we need the microscale for? In this view, the macroscale really does push around the microscale, and macroscale events really do cause microscale events.

However, it's important to admit that there are multiple ontological interpretations of what's going on in such cases of causal emergence. For there is another interpretation, which is that describing the system at any particular scale, whether macro or micro, is giving a low-dimensional slice of a high-dimensional object. In this view, systems don't have any one particular scale, they occur at multiple scales simultaneously. We, foolish hairless apes, draw out, say, the mechanisms of a system at a particular scale, and then think that this is the system. However, this does not necessarily lead us to relativism. For any given system has some subset of natural descriptions that each add something causally, and there may be ways to appropriately apportion the causation across scales. Imagine that the causal emergence of a given macroscale was 2 bits (as measured by effective information). But at the microscale, the effective information is only 1 bit, meaning at the macroscale it was 3 bits. We can say there are

3 bits of total causation, with 1 bit being down at the microscale, and 2 bits up at the macroscale. As you might have guessed, there are further details involved in any appropriating schema like this one.

But even if one is agnostic between these two metaphysical views, there is no interpretation wherein you yourself (or the macroscale most closely identified with you) are not a significant *contributory* cause of your actions in a way that's irreducible.

The exclusion argument is not the sole argument against free will, just behind a lot of the most popular claims that we lack it. However, I suspect that similar disproofs await for the remaining anti–free will arguments. To give a scientific definition of free will we must paint in negative space. Rather than hinging it on one argument, a rich definition of free will is based on a constellation of conditions, all of which modern science has given us reason to believe exist in physical systems—indeed, conditions that our brains themselves likely satisfy.

Bereitschaftspotential!

The name gives away that it was discovered by German scientists. The *Bereitschaftspotential* is the change in brain waves leading up to a voluntary action, a "potential" discovered originally in 1965.[5] The English term for it is the "readiness potential," and it's detectable by averaging brain activity before an action is taken. At the time of its discovery, it was posited to be indicatory of planning to initiate an action. The neuroscientist Benjamin Libet conducted a famous experiment that took advantage of this, one where he had patients watch a clock and remember when they made the decision to make a spontaneous movement. It turned out that the readiness potential started to build prior to when they consciously noted they had made a decision.[6] A huge amount has been made of this experiment. Does it disprove free will? Does consciousness still have a "veto" where it can decide to act contrary to the readiness potential?[7] And so on.

As with many of the big claims in neuroscience that show up in textbooks, recent research has called Libet's into doubt. Not that he engaged in malpractice or misrepresentation, but rather, using more modern techniques, it was shown to be much harder to tell when a movement would occur prior to the conscious decision to act, at least based on the readiness potential alone.[8] Regardless, Libet's supposed experimental refutation of free will is merely a specific example of an "argument from previous events" against free will. The argument goes like this: if it is the case that distant events in the past cause your actions, you aren't the real cause of your actions.[9] There's nothing special about the readiness potential in this regard, it's just a nice example.

Yet if causal influence comes in degrees, we can immediately see a problem with this. Just as the theory of causal emergence shows that causal influence might peak at a particular spatiotemporal scale, it's likely that causal influence can peak for recent events, rather than past events. As we move further into the past, the causal influence between past events and current events is less determinate and more degenerate, meaning that the distant past is a weaker cause than the recent past. This means that, even if distant events lead to current events, they are not nearly as strong (or contributory) a cause as recent events. The same can be true for judging if actions are actually caused based on factors internal to an agent or external.[10] Again, there are multiple ontological interpretations from this fact: Are recent events "screening off" past events so that distant events aren't causes at all, or can one work out an apportioning schema, or is it all relative? Regardless of which one you pick, the argument from previous events falls apart.

Occasionally the argument from previous events is framed in other ways, such as the "consequence argument," which rests on the idea that since you have no control over the laws of nature, or what went on before you were born, and since your actions are *solely* a result of these things, then you technically are not the cause of your actions.[11] But if, again, causation comes in degrees, and fades away with time, this doesn't seem like a big deal, since neither distant past events nor the laws of physics getting set at the beginning of the universe would be the strongest cause.

At this point, skeptics might argue that, putting aside all discussion of the causes of behavior, it's still problematic for free will that the readiness potential can be used to *predict* people's future behavior. Yet this view does not grapple with how our mathematical understanding of the limits of prediction has evolved in recent decades.

The Consolation of Chaos

Whiling away the time in prison near the end of the slow-motion fall of Rome and beginning of the Dark Ages, awaiting his eventual execution, Boethius, another Roman consul and philosopher, passed the time writing what would become one of the foundational classics of medieval literature: *The Consolation of Philosophy*.

Boethius had attempted to defend a fellow politician against charges of treason and ended up being lumped in with the traitor; he would eventually be tortured to death for his supposed crime. But before that, while in prison in the year 523 CE, he produced one of the leading intellectual texts of the Christian era in *The Consolation*. Grappling with an understandable depression, Boethius writes of trying to reconcile the omniscience of God with the idea of freedom of the will: "If from eternity He foreknows not only what men will do, but also their designs and purposes, there can be no freedom of the will."[12]

We can transpose God in this argument into a more contemporary notion of a scientist who has access to unlimited data and measurement capabilities, and is trying to predict what will happen to some particular system (which could be the universe as a whole). Let us assume instead that it's predictions of a man, Boethius, in his cell. Whether Boethius is predictable is actually more up in the air than it first appears, despite the assumed omniscience. First, as we've discussed, if there is actual noise down at the basement of physics that could "trickle up," then it would be impossible to predict from the current state the unique future occurrences. However, even in this case, God (or an omniscient scientist) could offer forth a set of probability distributions that would appropriately describe the future of Boethius. But let us set this complication aside, as it ends up being irrelevant, and briefly assume the most deterministic of universes, that is, those with only a single possible future.

To explain how the argument from prediction falls apart even in a deterministic universe, we must take a brief detour into what was going on in the 1980s during the heady days of establishing the scientific study of complexity and, with it, chaos theory—and with that, the very notion of prediction itself. What science discovered was that there are no shortcuts for predicting some systems—even if they are the result of deterministic laws that are themselves simple.

This discovery is easiest to talk about in terms of computer programs, where it is called "computational irreducibility." What it means is that for programs past a certain complexity, there is no shorter program that you can use to predict the results: the only thing you can do is run the program and observe what happens. In Wolfram's *A New Kind of Science* he defines this shift in how science thinks about prediction because of computational irreducibility thusly:

> In traditional science it has usually been assumed that if one can succeed in finding definite underlying rules for a system then this means that ultimately there will always be a fairly easy way to predict how the system will behave.... But now computational irreducibility leads to a much more fundamental problem with prediction. For it implies that even if in principle one has all the information one needs to work out how some particular system will behave, it can still take an irreducible amount of computational work actually to do this. Indeed, whenever computational irreducibility exists in a system it means that in effect there can be no way to predict how the system will behave except by going through almost as many steps of computation as the evolution of the system itself.... So what this means is that systems one uses to make predictions cannot be expected to do computations that are any more sophisticated than the computations that occur in all sorts of systems whose behavior we might try to predict. And from this it follows that for many systems no systematic prediction can be done, so that there is no general way to shortcut their process of evolution, and as a result their behavior must be considered computationally irreducible.[13]

Wolfram himself thinks this is related to free will, although he backs out at the last moment to instead frame computational irreducibility as merely a sufficient condition for our *perception* of others having free will. However, I think it offers much more than that, at least, combined with causal emergence.

For think on it. If a system in question (like a human in a cell) is indeed computationally irreducible, then God (or an omniscient scientist, if you prefer) is in an awkward situation. They can predict what will happen, but only by simulating the system in such perfect detail that it is identical, from a functional perspective, to the original one. So let us say that Boethius scratches his nose. God could predict this by recreating Boethius and the room down at the finest atomic level, and then letting Boethius II perform his action (turns out it's nose scratching), and then going back to the original Boethius and exclaiming (privately, or else it would influence the events), "Ha! You will scratch your nose in five seconds, looks like you're simply an automaton after all!" This seems very different in important ways from standard prediction of Boethius. First, the Boethius system seems to be predictable only via, effectively, a form of time travel. Meaning that Boethius II somehow has to be "faster" than Boethius I; that is, if we think of each as a computer program, the Turing machine running Boethius II must be faster than the one running Boethius I. Already this seems quite problematic (maybe you can't run physics any faster than physics goes). It is kind of like saying "I can predict what you will do, but only if I can time travel. If I can't time travel, I can't ever, no matter how many resources are at my disposal, actually predict what you will do." So the anti–free will advocate is forced to deflate the word "prediction" to mean merely observation plus time travel, and this certainly seems a much weaker threat to free will. Incredibly, this is basically Boethius's pro–free will argument, given fifteen hundred years ago, when he pointed out that for God, there is no difference in time between the future and the present—all events are simultaneous—effectively denying that God can time travel, since all times are as the present to him, so he can never be at one time alone, or move between them to compare!

But it gets worse for the free will denier here in a way that

Boethius did not foresee. For it appears that Boethius's clone, Boethius II, still has free will. To predict him, God would need a Boethius III. In other words, God appears to be caught in an infinite regress wherein he is constantly predicting human behavior at the cost of making humans he can't predict, merely observe, and then, after observing them, performing time travel in order to make use of this information to "predict" the original version. This is a definition of "prediction" that's been bled out completely. Whatever is going on here, it doesn't at all look like prediction; it looks instead like free will is an irreducible quality that you can never get rid of, only transfer to some new clone at a faster run-time.

The TV mini-series *Devs*, written and directed by Alexander Garland, illustrates this point well. The show's "MacGuffin" is a quantum computer that can simulate, in full detail, all of reality. This convinces the characters they don't have free will, especially once they use the machine to model themselves. In one scene, the cast of characters watches on a screen the same cast of characters, now perfectly modeled by the machine, except slightly ahead of them in time. The result is a horror where they scream for it to be turned off, since the actors on screen are ahead of them, screaming the same, and they react exactly the same, and their horrible realization is that they are merely mimics following inexorably along invisible train tracks.

But how could such a machine actually work? If you saw what you were going to do, wouldn't you do something different? The free-will denier must say that the future you are "following" is in turn being exposed to an even further future you, and so therefore your reactions must necessarily be the same, since the future you is seeing exactly what you see (someone ahead of them, and they in turn are the horrified mimics). Yet is this not an obvious impossibility? This would mean that the machine must be simulating itself, for in its predicted version of the future there must be another predicted version of the future, and so on, ad infinitum, spiraling out into an Escher-like infinity of impossibly demanding computation.

The Endless Bickering Around Fatalism

Now that they have been refuted as disproving free will, giving anti–free will arguments without any reference to prediction or causation turns out to actually be quite difficult. One possibility is to rely on the argument that our actions are fated via some sort of logical dependency, one that does not necessarily have to rely on prediction or causation (although it is easy to slip into that language, and one must watch oneself for that).

This form of "argument from fate" goes back all the way to Aristotle's *De Interpretatione*. However, the clearest example of logical fatalism about free will was given in a very famous philosophy paper by Richard Taylor.[14] He argued that a set of reasonable logical assumptions, like that all propositions are either true or false, entailed that free will was impossible.

It's been pointed out, however, that many of these assumptions are quite debatable.[15] If we assume an uncertain world (which would likely be entailed if scientific incompleteness is true, or if there is irreducible indeterminism in physics somewhere, etc.), it would be absurd to say that predictions about future states of affairs, like sea battles, are already true or false at the time they're made, and so logically entail the outcome. In an uncertain world, not all logical truth values (like predictions about future events) are defined at the time of their statements (indeed, Taylor himself advocated rejecting this assumption).

It's also worth noting that if logical entailment of the future from the past is not true, that seems equally problematic. If your behavior was not entailed by your brain state, this would be deeply concerning—it seems we want the cause of our caring for our child to be our love for them, or our estrangement from a friend to really be due to our anger at their actions, not due to an uncontrollable randomness. Therefore, we should already be skeptical that logical fatalism is a well-posed attack on free will, since its opposite, a universe where the future is not dependent on the past (not even randomly, as there is no probability distribution drawn from it), seems unappealing.

While no scientific discovery can impact how we interpret logic, the interpretation itself is also up for debate. This has been pointed out by an unlikely source: the writer David Foster Wallace. In his senior thesis at Amherst College, Wallace showed that the fatalism argument is based on a mere semantic conflation over different types of possibilities.[16]

As logical fatalism is the most philosophical and the most abstract argument against free will, it's therefore the most debatable. It is so complex, and so distant from science, that I can imagine the argument over it going on forever. At some point, we must leave the room of philosophers in armchairs behind.

What Is It to Be Free?

Putting aside the case of logical fatalism, the other conditions we've overviewed allow us to give a sketch of what a scientific definition of free will looks like. Having free will means being an agent that is causally emergent at the relevant level of description, for whom recent internal states are causally more relevant than distant past states, and who is computationally irreducible. We cannot know for sure, but it is very likely in my view, that the human brain satisfies these conditions, particularly if we do live in a fundamentally uncertain world. And while the conclusion of scientific incompleteness is speculative, it would entail that we do indeed live in such an uncertain world.

Notably, a great many thinkers have pointed to uncertainty as being somehow fundamental for free will, although an equal number of thinkers have responded by pointing out that uncertainty doesn't by itself necessitate free will. And it is true that uncertainty alone does nothing for establishing free will. But as we've seen, uncertainty does enable the *conditions* for a set of properties, like being causally emergent and computationally irreducible, which, when put together, seem to satisfy an intuitive definition of free will.

Science is not always a universal acid. Sometimes it supports our most cherished ideas, rather than demolish them. It picks us up, rather than knocking us down. Call it the consolation of science. We may be hairless apes, but we are conscious, and that is indeed something special and unique, as the paradoxes around it attest to. Studying consciousness scientifically requires exploring the hybrid zone where the qualitative meets the quantitative, a unique metaphysical ecosystem. And it is possible that this zone will never be resolved to our satisfaction in the way other fields of science are, that it, and therefore we, will always remain paradoxical, mysterious as a deep-sea trench.

Nor is the scientific view of the world, even if it is necessarily incomplete, a universally reductive one. A deeper understanding of causation, prediction, and computation not only gives us a disproof of universal reduction, but a notion of emergence, and with it, a conception of free will that's truly worth having.

Perhaps it is fitting for a book on the mixing of the extrinsic and intrinsic perspectives that I cannot hide my own subjectivity here. The author must appear and confess. For I find the resultant freedom dizzying and heady. It is, in a way, a long-sought-after historical dream: a full-throated freedom that is not retreating from the scientific worldview but firmly based in its discoveries. For so long, writers, politicians, poets, saints, mystics, thinkers of all types and creeds, have wondered over the fault in our stars. We can hear their hope for freedom in Percy Shelley's "Ode to Liberty!" and we can hear it in the singing caged bird of Maya Angelou; we can hear it in the cry of Moses for the Pharaohs to let his people go. And we hear it not just in the great revolutions and statements of history, but in the quiet of everyday mundane life, in the masses of humanity that are just as real and full of introspection and curiosity as the famous names we're all famil- iar with. We can hear it in the small acts of private individuals; we can hear it when they decide between cereal brands in the supermarket, when they pick a particular book from a shelf; we can hear it in whom they choose to marry, in how they decide to treat their children. We can hear it in William James's closing of his diary and choosing to exit his room and take a walk in the morning light.

Acknowledgments

The greatest thank-you, really an impossibly great one, goes to my wife, who accompanied me on this journey from start to finish. Thanks also to first readers who gave feedback on parts of the manuscript, especially Anil Seth, Hedda Hassel Mørch, Brennan Klein, and Thomas Varley. Additionally, a special thanks to Alex Criddle for helping research Chapter 2, as well as Liberty Severs, who assisted with research for Chapter 4. Thanks as well to my agent, Susan Golomb, who encouraged me to work on this book in particular out of all the projects I was considering, as well as Ben Loehnen, my editor at Simon & Schuster, who patiently helped the book become all it could be.

Notes

CHAPTER 1: HUMANITY'S TWO PERSPECTIVES ON THE WORLD

1. David Chalmers, *The Conscious Mind: In Search of a Fundamental Theory* (Oxford: Oxford University Press, 1997).
2. Christof Koch, *Consciousness: Confessions of a Romantic Reductionist* (Cambridge, MA: MIT Press, 2012).
3. Masafumi Oizumi, Larissa Albantakis, and Giulio Tononi, "From the Phenomenology to the Mechanisms of Consciousness: Integrated Information Theory 3.0," *PLoS Computational Biology* 10, no. 5 (2014): e1003588.

CHAPTER 2: THE DEVELOPMENT OF
THE INTRINSIC PERSPECTIVE

1. Julian Jaynes, *The Origin of Consciousness in the Breakdown of the Bicameral Mind* (Boston: Houghton Mifflin Harcourt, 1976), 69.
2. Weichen Song et al., "A Selection Pressure Landscape for 870 Human Polygenic Traits," *Nature Human Behaviour* 5, no. 12 (2021): 1731–1743.
3. Ned Block, "Review of Julian Jaynes, Origin of Consciousness in the Breakdown of the Bicameral Mind," *Boston Globe*, March 6, 1977, https://philpapers.org/archive/BLOROJ-2.pdf.
4. Bill Rowe, "Retrospective: Julian Jaynes and the Origin of Consciousness in the Breakdown of the Bicameral Mind," *American Journal of Psychology* 125, no. 3 (2012): 369–381.

Notes

5. Scott Alexander, "Book Review: The Origin of Consciousness in the Breakdown of the Bicameral Mind," *Slate Star Codex*, June 1, 2020, https://slatestarcodex.com/2020/06/01/book-review-origin-of-consciousness-in-the-breakdown-of-the-bicameral-mind/.

6. James W. Moore, "'They Were Noble Automatons Who Knew Not What They Did': Volition in Jaynes' The Origin of Consciousness in the Breakdown of the Bicameral Mind," *Frontiers in Psychology* 12 (2021).

7. Heinz Wimmer and Josef Perner, "Beliefs About Beliefs: Representation and Constraining Function of Wrong Beliefs in Young Children's Understanding of Deception," *Cognition* 13, no. 1 (1983): 103–128.

8. Rowe, "Retrospective: Julian Jaynes."

9. Bruno Snell, *The Discovery of the Mind*, trans. Thomas G. Rosenmeyer (Cambridge, MA: Harvard University Press, 1953).

10. Thomas Nagel, *The View from Nowhere* (Oxford: Oxford University Press, 1989).

11. A. Erman, *The Literature of the Ancient Egyptians: Poems, Narratives, and Manuals of Instruction from the Third and Second Millenia B.C.* (New York: Benjamin Blom, Inc., 1927), 109.

12. G. Robins, *Proportion and Style in Ancient Egyptian Art* (Austin: University of Texas Press, 1994).

13. G. Richter, *Art and Human Consciousness* (Hudson, NY: Steiner Books, 1985), 126.

14. W. Simpson, W. (ed.), *The Literature of Ancient Egypt* (New Haven, CT: Yale University Press, 2003), 179.

15. M. Lichtheim (ed.), *Ancient Egyptian Literature, Volume I: The Old and Middle Kingdoms* (Berkeley: University of California Press, 1973), 92.

16. Ibid., 17.

17. Ibid., 18–27.

18. R. I. M. Dunbar, *Grooming, Gossip, and the Evolution of Language* (Cambridge, MA: Harvard University Press, 1998).

19. N. Block, "On a Confusion About a Function of Consciousness," *Behavioral and Brain Sciences* 18, no. 2 (1995): 227–247.
20. A. Erman, *Life in Ancient Egypt* (London, 1894), 389.
21. M. Lichtheim (ed.), *Ancient Egyptian Literature, Volume II: The New Kingdom* (Berkeley: University of California Press, 2006), 192.
22. Jaynes, *Origin of Consciousness*, 72.
23. Ibid., 272.
24. G. Lessing, *Laocoon: An Essay on the Limits of Painting and Poetry* (1766), xv.
25. F. Yates, *The Art of Memory* (London: Routledge and Kegan Paul, 1966), 1–3.
26. J. Foer, *Moonwalking with Einstein: The Art and Science of Remembering Everything* (New York: Penguin, 2012).
27. Snell, *Discovery of the Mind*, 125.
28. Ibid., 250.
29. Cicero, in *The Letters of Cicero: The Whole Extant Correspondence in Chronological Order, in Four Volumes*, trans. Evelyn Shirley Shuckburgh (London: Bell and Sons, 1908), 207.
30. Yates, *The Art of Memory*, 4.
31. J. Atkins, "Euripides's Orestes and the Concept of Conscience in Greek Philosophy," *Journal of the History of Ideas* 75, no. 1 (2014): 1–22.
32. C. Valerius Catullus, "Poem 85," *Carmina*, trans. Leonard C. Smithers (Perseus Project).
33. O. Brockett and F. Hildy, *History of Theater* (Boston: Pearson Education Limited, 2014).
34. C. Wickham, *The Inheritance of Rome: Illuminating the Dark Ages 400–1000* (New York: Penguin Books, 2010).
35. B. Ward-Perkins, *The Fall of Rome and the End of Civilisation* (Oxford: Oxford University Press, 2006), 164.
36. Kristina Milnor, *Graffiti and the Literary Landscape in Roman Pompeii* (Oxford: Oxford University Press, 2014).
37. Lockett, "Embodiment, Metaphor, and the Mind in Old English Narrative," in *The Emergence of Mind: Representations of Conscious-*

ness in Narrative Discourse in English, ed. D. Herman (Lincoln: University of Nebraska Press, 2011).

38. J. Sedivy, "Why Doesn't Ancient Fiction Talk About Feelings?," *Nautilus* 47 (2017).

39. E. Gardiner, *Visions of Heaven and Hell Before Dante* (New York: Italica Press, 1989), 57–65.

40. L. Zunshine, *Why We Read Fiction: Theory of Mind and the Novel* (Columbus: Ohio State University Press, 2006).

41. M. Cervantes, *Don Quixote*, trans. John Ormsby (London: Smith, Elder, 1885), 1–33.

42. George Eliot, *Middlemarch*, in D. Herman (ed.), *The Emergence of Mind: Representations of Consciousness in Narrative Discourse in English* (Lincoln: University of Nebraska Press, 2011), 26.

43. I. Watt, *The Rise of the Novel* (Berkeley: University of California Press, 2001).

44. V. Woolf, "Modern Fiction," in *The Common Reader* (New York: Harcourt, Brace & Company, 1925), 150–158.

45. Zunshine, *Why We Read Fiction*.

46. D. Cohn, *Transparent Minds: Narrative Modes for Presenting Consciousness in Fiction* (Princeton, NJ: Princeton University Press, 1978).

47. A. Palmer, *Social Minds in the Novel* (Columbus: Ohio State University Press, 2010).

48. T. Wolfe, *Hooking Up* (New York: Macmillan, 2000), 169.

49. E. P. Hoel, "Fiction in the Age of Screens," *New Atlantis* (2016), 93–109.

50. X. Chen et al., "The Role of Personality and Subjective Exposure Experiences in Posttraumatic Stress Disorder and Depression Symptoms Among Children Following Wenchuan Earthquake," *Scientific Reports* 7, no. 1 (2017): 1–9.

51. A. Russell (ed.), "Autographs and Signatures," in *The Guinness Book of Records 1987* (Stamford, CT: Guinness Books, 1986).

52. F. Petrie, *Egyptian Tales, Volume 1: Translated from the Papyri* (London: Methuen & Co., 1899), 81–92.

53. Homer, *The Odyssey with an English Translation by A. T. Murray, Ph.D.* (Cambridge, MA: Harvard University Press, 1919), 250–258.

54. J. Joyce, *Ulysses* (Ware, UK: Wordsworth Editions, 2010), 283.

CHAPTER 3: THE DEVELOPMENT OF
THE EXTRINSIC PERSPECTIVE

1. Henri Poincaré, *Science and Hypothesis* (New York: The Science Press, 1905).

2. Arthur Miller, "Henri Poincaré: The Unlikely Link Between Einstein and Picasso," *Guardian*, July 17, 2012, https://www.theguardian.com/science/blog/2012/jul/17/henri-poincare-einstein-picasso.

3. Rajarshi Ghosh, "Book Review: Einstein, Picasso: Space, Time and the Beauty That Causes Havoc," *Einstein Quarterly Journal of Biology and Medicine* 19 (2002): 45–46.

4. David C. Lindberg, *The Beginnings of Western Science: The European Scientific Tradition in Philosophical, Religious, and Institutional Context, Prehistory to AD 1450* (Chicago: University of Chicago Press, 2007).

5. Will Durant, *Our Oriental Heritage: The Story of Civilization, Volume I* (New York: Simon & Schuster, 1954), 179.

6. Lindberg, *The Beginnings of Western Science*.

7. Bertrand Russell, *History of Western Philosophy* (New York: Simon & Schuster, 2007).

8. Peter Dear, *Revolutionizing the Sciences: European Knowledge and Its Ambitions, 1500–1700* (Princeton, NJ: Princeton University Press, 2001).

9. Sylvia Berryman, "Democritus," in *The Stanford Encyclopedia of Philosophy*, December 2, 2016, https://plato.stanford.edu/archives/win2016/entries/democritus/.

10. Dear, *Revolutionizing the Sciences*.

11. M. de Voltaire, *The Works of Voltaire, Vol. XII (Age of Louis XIV)* (New York: E. R. DuMont, 1751), https://oll.libertyfund.org/title/fleming-the-works-of-voltaire-vol-xii-age-of-louis-xiv.

12. Joel Mokyr, *A Culture of Growth: The Origins of the Modern Economy* (Princeton, NJ: Princeton University Press, 2016).

13. Dena Goodman, "Enlightenment Salons: The Convergence of Female and Philosophic Ambitions," *Eighteenth-Century Studies* 22, no. 3 (1989): 329–350, https://doi.org/10.2307/2738891.

14. David Graeber and David Wengrow, *The Dawn of Everything: A New History of Humanity* (London: Penguin UK, 2021).

15. Sean Carroll, "Beyond Falsifiability: Normal Science in a Multiverse," in *Why Trust a Theory? Epistemology of Fundamental Physics*, ed. Radin Dardashti, Richard Dawid, and Karim Thébault (Cambridge, UK: Cambridge University Press, 2019), 300–314.

16. Galileo Galilei, "The Assayer," in *Discoveries and Opinions of Galileo*, trans. Stillman Drake (New York: Doubleday & Co., 1957), 237–238, https://www.princeton.edu/~hos/h291/assayer.htm.

17. Philip Goff, *Galileo's Error: Foundations for a New Science of Consciousness* (New York: Pantheon Books, 2019).

18. B. L. Keeley, "The Early History of the Quale and Its Relation to the Senses," in *The Routledge Companion to Philosophy of Psychology* (London: Routledge, 2019), 71–89.

19. David Chalmers, *The Conscious Mind: In Search of a Theory of Conscious Experience* (Oxford: Oxford University Press, 1996).

20. Goff, *Galileo's Error*, 21.

CHAPTER 4: NEUROSCIENCE IN NEED OF A REVOLUTION

1. John Mark Taylor, "Mirror Neurons After a Quarter Century: New Light, New Cracks," *Harvard University* blog, July 25, 2016, https://sitn.hms.harvard.edu/flash/2016/mirror-neurons-quarter-century-new-light-new-cracks/.

2. G. di Pellegrino et al., "Understanding Motor Events: A Neurophysiological Study," *Experimental Brain Research* 91, no. 1 (1992): 176–180.

3. Vilayanur Ramachandran, "Mirror Neurons and Imitation Learning as the Driving Force Behind the Great Leap Forward in Human

Evolution," *Edge*, May 31, 2000 https://www.edge.org/conversation/vilayanur_ramachandran-mirror-neurons-and-imitation-learning-as-the-driving-force.

4. Vittorio Caggiano et al., "View-Based Encoding of Actions in Mirror Neurons of Area F5 in Macaque Premotor Cortex," *Current Biology* 21, no. 2 (2011): 144–148.

5. Vittorio Caggiano et al., "Mirror Neurons Encode the Subjective Value of an Observed Action," *Proceedings of the National Academy of Sciences* 109, no. 29 (2012): 11848–11853.

6. James M. Kilner and Roger N. Lemon, "What We Know Currently About Mirror Neurons,"*Current Biology* 23 (2013): R1057–R1062.

7. Lindsay M. Oberman, Jaime A. Pineda, and Vilayanur S. Ramachandran, "The Human Mirror Neuron System: A Link Between Action Observation and Social Skills," *Social Cognitive and Affective Neuroscience* 2, no. 1 (2007): 62–66.

8. Suresh D. Muthukumaraswamy and Blake W. Johnson, "Primary Motor Cortex Activation During Action Observation Revealed by Wavelet Analysis of the EEG," *Clinical Neurophysiology* 115, no. 8 (2004): 1760–1766.

9. Riccardo Viaro et al., "Neurons of Rat Motor Cortex Become Active During Both Grasping Execution and Grasping Observation," *Current Biology* 31, no. 19 (2021): 4405–4412.

10. Maria Carrillo et al., "Emotional Mirror Neurons in the Rat's Anterior Cingulate Cortex," *Current Biology* 29, no. 8 (2019): 1301–1312.

11. Richard Mooney, "Auditory-Vocal Mirroring in Songbirds, "*Philosophical Transactions of the Royal Society B: Biological Sciences* 369, no. 1644 (2014): 20130179.

12. Lindsay M. Oberman,and Vilayanur S. Ramachandran, "The Simulating Social Mind: The Role of the Mirror Neuron System and Simulation in the Social and Communicative Deficits of Autism Spectrum Disorders," *Psychological Bulletin* 133, no. 2 (2007): 310.

13. Antonia F. de C. Hamilton, "Reflecting on the Mirror Neuron System in Autism: A Systematic Review of Current Theories," *Developmental Cognitive Neuroscience* 3 (2013): 91–105.

14. Ilan Dinstein et al., "Normal Movement Selectivity in Autism," *Neuron* 66, no. 3 (2010): 461–469.

15. Gregory Hickok, *The Myth of Mirror Neurons: The Real Neuroscience of Communication and Cognition* (New York: W. W. Norton & Company, 2014).

16. Richard Cook and Geoffrey Bird, "Do Mirror Neurons Really Mirror and Do They Really Code for Action Goals?," *Cortex* 49, no. 10 (2013): 2944–2945.

17. Joanna Moncrieff et al., "The Serotonin Theory of Depression: A Systematic Umbrella Review of the Evidence," *Molecular Psychiatry* (2022): 1–14.

18. Daniel Cressey, "Psychopharmacology in Crisis,"*Nature*, June 14, 2011, https://doi.org/10.1038/news.2011.367.

19. Ned Pagliarulo, "Pfizer Pulls Back from Neuroscience, Ending Research," *BioPharma Dive*, https://www.biopharmadive.com/news/pfizer-ends-neuroscience-alzheimers-research-cuts-300-jobs/514210/.

20. Andrew Dunn, "Amgen Exits Neuroscience R&D as Pharma Pulls Back from the Field," *BioPharma Dive*, October 30, 2019, https://www.biopharmadive.com/news/amgen-exits-neuroscience-rd-as-pharma-pulls-back-from-field/566157/.

21. Jacob Bell, "Big Pharma Backed Away from Brain Drugs. Is a Return in Sight?" *BioPharma Dive*, January 29, 2020, https://www.biopharmadive.com/news/pharma-neuroscience-retreat-return-brain-drugs/570250/.

22. Klaus-Peter Lesch et al., "Association of Anxiety-Related Traits with a Polymorphism in the Serotonin Transporter Gene Regulatory Region," *Science* 274, no. 5292 (1996): 1527–1531.

23. David Dobbs, "The Science of Success," *Atlantic*, December 2009, https://www.theatlantic.com/magazine/archive/2009/12/the-science-of-success/307761/.

24. Dave Davies, "Is Your Child an Orchid or a Dandelion? Unlocking the Science of Sensitive Kids," NPR, March 4, 2019, https://www.npr.org/sections/health-shots/2019/03/04/699979387/is

-your-child-an-orchid-or-a-dandelion-unlocking-the-science-of
-sensitive-kids?t=1647970124742.

25. Richard Border et al., "No Support for Historical Candidate Gene or Candidate Gene-By-Interaction Hypotheses for Major Depression Across Multiple Large Samples," *American Journal of Psychiatry* 176, no. 5 (2019): 376–387.

26. Scott Alexander, "5-HTTLPR: A Pointed Review," *Slate Star Codex*, May 5, 2019, https://slatestarcodex.com/2019/05/07/5-httlpr-a -pointed-review/.

27. John P. A. Ioannidis, "Why Most Published Research Findings Are False," *PLOS Medicine* 2, no. 8 (2005): e124, https://doi.org/10.1371 /journal.pmed.0020124.

28. C. Glenn Begley and Lee M. Ellis, "Raise Standards for Preclinical Cancer Research,"*Nature* 483, no. 7391 (2012): 531–533.

29. Timothy M. Errington et al., "Investigating the Replicability of Preclinical Cancer Biology," *Elife* 10 (2021): e71601.

30. Open Science Collaboration, "Estimating the Reproducibility of Psychological Science,"*Science* 349, no. 6251 (2015): aac4716.

31. Colin F. Camerer et al., "Evaluating the Replicability of Social Science Experiments in Nature and Science Between 2010 and 2015," *Nature Human Behaviour* 2, no. 9 (2018): 637–644.

32. Roy F. Baumeister, Ellen Bratslavsky, Mark Muraven, and Dianne M. Tice, "Ego Depletion: Is the Active Self a Limited Resource?," in *Self-Regulation and Self-Control* (London: Routledge, 2018), 6–44.

33. Kathleen D. Vohs, Nicole L. Mead, and Miranda R. Goode, "The Psychological Consequences of Money," *Science* 314, no. 5802 (2006): 1154–1156.

34. Philip G. Zimbardo, "The Mind Is a Formidable Jailer: A Pirandellian Prison," *New York Times*, April 8, 1973, https://www.nytimes .com/1973/04/08/archives/a-pirandellian-prison-the-mind-is-a -formidable-jailer.html.

35. Thibault Le Texier, "Debunking the Stanford Prison Experiment," *American Psychologist* 74, no. 7 (2019): 823.

36. Doug Rohrer, Harold Pashler, and Christine R. Harris, "Discrep-

ant Data and Improbable Results: An Examination of Vohs, Mead, and Goode (2006)," *Basic and Applied Social Psychology* 41, no. 4 (2019): 263–271.

37. Miguel A. Vadillo, "Ego Depletion May Disappear by 2020," *Social Psychology* 50, no. 5–6 (2019): 282.

38. Justin Kruger and David Dunning, "Unskilled and Unaware of It: How Difficulties in Recognizing One's Own Incompetence Lead to Inflated Self-Assessments," *Journal of Personality and Social Psychology* 77, no. 6 (1999): 1121.

39. Edward Nuhfer et al., "How Random Noise and a Graphical Convention Subverted Behavioral Scientists' Explanations of Self-Assessment Data: Numeracy Underlies Better Alternatives," *Numeracy: Advancing Education in Quantitative Literacy* 10, no. 1 (2017).

40. Gilles E. Gignac and Marcin Zajenkowski, "The Dunning-Kruger Effect Is (Mostly) a Statistical Artefact: Valid Approaches to Testing the Hypothesis with Individual Differences Data," *Intelligence* 80 (2020): 101449.

41. Gang Chen, Paul A. Taylor, and Robert W. Cox, "Is the Statistic Value All We Should Care About in Neuroimaging?," *Neuroimage* 147 (2017): 952–959.

42. Craig M. Bennett, Michael B. Miller, and George L. Wolford, "Neural Correlates of Interspecies Perspective Taking in the Post-Mortem Atlantic Salmon: An Argument for Multiple Comparisons Correction," *Neuroimage* 47, Supplement 1 (2009): S125.

43. A. Tlaie et al., "Does the Brain Care About Averages? A Simple Test," *bioRxiv* (2022), 1.

44. Marc-Andre Schulz et al, "Performance Reserves in Brain-Imaging-Based Phenotype Prediction," *bioRxiv* (2022).

45. Scott Marek et al., "Reproducible Brain-Wide Association Studies Require Thousands of Individuals," *Nature* 603, no. 7902 (2022): 654–660.

46. Dan D. Stettler et al., "Axons and Synaptic Boutons Are Highly Dynamic in Adult Visual Cortex," *Neuron* 49, no. 6 (2006): 877–887.

47. Rodrigo Quian Quiroga, Itzhak Fried, and Christof Koch, "Brain

Cells for Grandmother," *Scientific American* 308, no. 2 (2013): 30–35.

48. Daniel Deitch, Alon Rubin, and Yaniv Ziv, "Representational Drift in the Mouse Visual Cortex," *Current Biology* 31, no. 19 (2021): 4327–4339.

49. Kyle Aitken, Marina Garrett, Shawn Olsen, and Stefan Mihalas, "The Geometry of Representational Drift in Natural and Artificial Neural Networks," *PLOS Computational Biology* 18, no. 11 (2022): e1010716.

50. Andrew Gelman and Eric Loken, "The Garden of Forking Paths: Why Multiple Comparisons Can Be a Problem, Even When There is No 'Fishing Expedition' or 'p-hacking' and the Research Hypothesis Was Posited Ahead of Time," *Department of Statistics, Columbia University* 348 (2013): 1–17.

51. Dorothy Bishop, "Rein in the Four Horsemen of Irreproducibility," *Nature* 568, no. 7753 (2019).

52. Eric Jonas and Konrad Paul Kording, "Could a Neuroscientist Understand a Microprocessor?," *PLoS Computational Biology*13, no. 1 (2017): e1005268.

53. Davide Castelvecchi, "Can We Open the Black Box of AI?," *Nature News* 538, no. 7623 (2016): 20.

54. Stephen Thornton, "Karl Popper," in *The Stanford Encyclopedia of Philosophy* (Winter 2022 Edition), September 12, 2022, https://plato.stanford.edu/archives/win2022/entries/popper.

55. Thomas Kuhn, *The Structure of Scientific Revolutions* (Chicago: University of Chicago Press, 1962).

56. Imre Lakatos, "Falsification and the Methodology of Scientific Research Programs," in *Criticism and the Growth of Knowledge*, ed. Imre Lakatos and Alan Musgrave (Cambridge, UK: Cambridge University Press, 1970), 91–196.

57. Alan Musgrave and Charles Pigden, "Imre Lakatos," *The Stanford Encyclopedia of Philosophy* (Summer 2021 Edition), April 26, 2021, https://plato.stanford.edu/archives/sum2021/entries/lakatos.

58. Kuhn, *The Structure of Scientific Revolutions*, 67.

59. Christopher W. Tyler, "Peripheral Color Demo," *i-Perception* 6, no. 6 (2015): 2041669515613671.

60. Denis G. Pelli and Katharine A. Tillman, "The Uncrowded Window of Object Recognition," *Nature Neuroscience* 11, no. 10 (2008): 1129–1135.

61. Lisandro N. Kaunitz, Elise G. Rowe, and Naotsugu Tsuchiya, "Large Capacity of Conscious Access for Incidental Memories in Natural Scenes," *Psychological Science* 27, no. 9 (2016): 1266–1277.

62. Catherine Tallon-Baudry, "The Topological Space of Subjective Experience," *Trends in Cognitive Sciences* 26, no. 12 (2022): 1068–1069.

63. Andrew Haun and Giulio Tononi, "Why Does Space Feel the Way It Does? Towards a Principled Account of Spatial Experience," *Entropy* 21, no. 12 (2019): 1160.

64. Giulio Tononi, "An Information Integration Theory of Consciousness," *BMC Neuroscience* 5, no. 1 (2004): 1–22.

65. Theodosius Dobzhansky, "Nothing in Biology Makes Sense Except in the Light of Evolution," *American Biology Teacher* 75, no. 2 (2013): 87–91.

CHAPTER 5: THE TWO HOUSES OF CONSCIOUSNESS RESEARCH

1. Francis Crick, *The Astonishing Hypothesis: The Scientific Search for the Soul* (New York: Charles Scribner's Sons, 1994).

2. Francis Crick and Christof Koch, "Towards a Neurobiological Theory of Consciousness," *Seminars in the Neurosciences* 2 (1990): 263–275.

3. Gerald M. Edelman, *Neural Darwinism: The Theory of Neuronal Group Selection* (New York: Basic Books, 1987).

4. Gerald M. Edelman, *The Remembered Present: A Biological Theory of Consciousness* (New York: Basic Books, 1989).

5. Francis Crick, "Neural Edelmanism," *Trends in Neurosciences* 12, no. 7 (1989): 240–248.

6. Naotsugu Tsuchiya, Melanie Wilke, Stefan Frässle, and Victor F.

Lamme, "No-Report Paradigms: Extracting the True Neural Correlates of Consciousness," *Trends in Cognitive Sciences* 19, no. 12 (2015): 757–770.

7. Hakwan C. Lau and Richard E. Passingham, "Relative Blindsight in Normal Observers and the Neural Correlate of Visual Consciousness," *Proceedings of the National Academy of Sciences* 103, no. 49 (2006): 18763–18768.

8. James D. Watson and Francis Crick, "Molecular Structure of Nucleic Acids: A Structure for Deoxyribose Nucleic Acid," *Nature* 171, no. 4356 (1953): 737–738.

9. Michael Wenzel et al., "Reduced Repertoire of Cortical Microstates and Neuronal Ensembles in Medically Induced Loss of Consciousness," *Cell Systems* 8, no. 5 (2019): 467–474.

10. Simone Sarasso et al., "Consciousness and Complexity: A Consilience of Evidence," *Neuroscience of Consciousness* 7, no. 2 (2021): 1–24.

11. Itay Yaron, Lucia Melloni, Michael Pitts, and Liad Mudrik, "The ConTraSt Database for Analysing and Comparing Empirical Studies of Consciousness Theories," *Nature Human Behaviour* 6, no. 4 (2022): 593–604.

12. Anil K. Seth, "Darwin's Neuroscientist: Gerald M. Edelman, 1929–2014," *Frontiers in Psychology* 5 (2014): 896.

13. David Hellerstein, "Plotting a Theory of the Brain," *New York Times*, May 22, 1988, https://www.nytimes.com/1988/05/22/magazine/plotting-a-theory-of-the-brain.html.

14. Johannes Kleiner and Erik Hoel, "Falsification and Consciousness," *Neuroscience of Consciousness* 2021, no. 1 (2021): niab001.

15. Bernard J. Baars, *A Cognitive Theory of Consciousness* (Cambridge, UK: Cambridge University Press, 1989).

16. Stanislas Dehaene, Michel Kerszberg, and Jean-Pierre Changeux, "A Neuronal Model of a Global Workspace in Effortful Cognitive Tasks," *Proceedings of the National Academy of Sciences* 95, no. 24 (1998): 14529–14534.

17. Lenore Blum and Manuel Blum, "A Theory of Consciousness from a

Theoretical Computer Science Perspective: Insights from the Conscious Turing Machine," *Proceedings of the National Academy of Sciences* 119, no. 21 (2022): e2115934119.

18. Giulio Tononi, "An Information Integration Theory of Consciousness," *BMC Neuroscience* 5, no. 1 (2004): 1–22.

19. Carl Zimmer, "Sizing Up Consciousness by Its Bits," *New York Times*, September 20, 2010, https://www.nytimes.com/2010/09/21/science/21consciousness.html.

20. Adenauer G. Casali et al., "A Theoretically Based Index of Consciousness Independent of Sensory Processing and Behavior," *Science Translational Medicine* 5, no. 198 (2013): 198ra105–198ra105.

21. Evan S. Lutkenhoff et al., "Subcortical Atrophy Correlates with the Perturbational Complexity Index in Patients with Disorders of Consciousness," *Brain Stimulation* 13, no. 5 (2020): 1426–1435.

22. Catherine Duclos et al., "Brain Responses to Propofol in Advance of Recovery from Coma and Disorders of Consciousness: A Preliminary Study," *American Journal of Respiratory and Critical Care Medicine* 205, no. 2 (2022): 171–182.

CHAPTER 6: PHENOMENOLOGICAL
THEORIES OF CONSCIOUSNESS

1. Louisa S. Cook, *Geometrical Psychology, Or, The Science of Representation: An Abstract of the Theories and Diagrams of BW Betts* (George Redway, 1887), https://ia903404.us.archive.org/31/items/geometricalpsych00cook/geometricalpsych00cook.pdf.

2. Ibid., 9.

3. Ibid., 12.

4. Ibid., 92.

5. Morton Hunt, *The Story of Psychology* (New York: Anchor, 2007), 147.

6. Edward Bradford Titchener, *Lectures on the Elementary Psychology of Feeling and Attention* (London: Macmillan, 1908).

7. Cook, *Geometrical Psychology*, 15.

8. Ibid.

9. Tononi, "An Information Integration Theory of Consciousness."

10. Giulio Tononi and Gerald M. Edelman, "Consciousness and Complexity," *Science* 282, no. 5395 (1998): 1846–1851.

11. Giulio Tononi, "Consciousness as Integrated Information: A Provisional Manifesto," *Biological Bulletin* 215, no. 3 (2008): 216–242.

12. Erik Hoel, *The World Makers*, unpublished.

13. Max Tegmark, "Improved Measures of Integrated Information," *PLoS Computational Biology* 12, no. 11 (2016): e1005123.

14. Matteo Grasso, Larissa Albantakis, Jonathan P. Lang, and Giulio Tononi, "Causal Reductionism and Causal Structures," *Nature Neuroscience* 24, no. 10 (2021): 1348–1355.

15. Sabine Hossenfelder, *Existential Physics: A Scientist's Guide to Life's Biggest Questions* (New York: Viking, 2022), 82.

16. William James, *The Principles of Psychology Volume I* (New York: Henry Holt, 1918), 160, https://mindsplain.com/wp-content/uploads/2020/08/The-Principles-of-Psychology-I-by-William-James_.pdf.

17. Elizaveta Levina and Peter Bickel, "The Earth Mover's Distance Is the Mallows Distance: Some Insights from Statistics," *Proceedings Eighth IEEE International Conference on Computer Vision, ICCV 2001* 2 (2001): 251–256.

18. Cook, *Geometrical Psychology*, 16.

19. Daniel C. Dennett, *Consciousness Explained* (Boston: Back Bay Books, 1991).

20. Herman Melville, *Moby-Dick; Or, The Whale* (New York: Harper & Brothers, 1851), 369, https://en.wikisource.org/wiki/Moby-Dick_(1851)_US_edition.

21. Matt Visser, "Acoustic Black Holes: Horizons, Ergospheres and Hawking Radiation," *Classical and Quantum Gravity* 15, no. 6 (1998): 1767.

22. Melvyn A. Goodale and A. David Milner, "Separate Visual Pathways for Perception and Action," *Trends in Neurosciences* 15, no. 1 (1992): 20–25.

23. Tim Bayne, "On the Axiomatic Foundations of the Integrated Information Theory of Consciousness," *Neuroscience of Consciousness* 2018, no. 1 (2018): niy007, 2.

24. Larissa Albantakis, Leonardo Barbosa, Graham Findlay, Matteo Grasso, Andrew M. Haun, William Marshall, William GP Mayner et al. "Integrated information (IIT) 4.0: Formulating the properties of phenomenal existence in physical terms." arXiv preprint arXiv:2211.14787 (2022).

25. Sean Carroll, "Beyond Falsifiability: Normal Science in a Multiverse," in *Why Trust a Theory? Epistemology of Fundamental Physics*, ed. Radin Dardashti, Richard Dawid, and Karim Thébault (Cambridge, UK: Cambridge University Press, 2019), 300–314.

26. Bernard J. Baars, "Global Workspace Theory of Consciousness: Toward a Cognitive Neuroscience of Human Experience," *Progress in Brain Research* 150 (2005): 45–53.

27. Kleiner and Hoel, "Falsification and Consciousness."

28. Adrien Doerig, Aaron Schurger, Kathryn Hess, and Michael H. Herzog, "The Unfolding Argument: Why IIT and other Causal Structure Theories Cannot Explain Consciousness," *Consciousness and Cognition* 72 (2019): 49–59.

29. Anton Maximilian Schäfer and Hans Georg Zimmermann, "Recurrent Neural Networks Are Universal Approximators," in *International Conference on Artificial Neural Networks* (Berlin: Springer, 2006), 632–640.

30. Kurt Hornik, Maxwell Stinchcombe, and Halbert White, "Multilayer Feedforward Networks Are Universal Approximators," *Neural Networks* 2, no. 5 (1989): 359–366.

31. Jake R. Hanson and Sara I. Walker, "Integrated Information Theory and Isomorphic Feed-Forward Philosophical Zombies," *Entropy* 21, no. 11 (2019): 1073.

32. Johannes Kleiner, "Brain States Matter: A Reply to the Unfolding Argument," *Consciousness and Cognition* 85 (2020): 102981.

33. Kleiner and Hoel, "Falsification and Consciousness."

34. Alan Turing, "On Computable Numbers, with an Application to the

Entscheidungsproblem: A Correction," *Proceedings of the London Mathematical Society* 2, no. 1 (1938): 544–546.

35. Stephen Wolfram, "Cellular Automata as Models of Complexity," *Nature* 311, no. 5985 (1984): 419–424.

36. Christof Koch, *The Feeling of Life Itself: Why Consciousness Is Widespread But Can't Be Computed* (Cambridge, MA: MIT Press, 2019).

37. Marcus Hutter, "A Gentle Introduction to the Universal Algorithmic Agent {AIXI}," in *Artificial General Intelligence*, ed. Ben Goertzel and Cassio Pennachin (Berlin: Springer, 2003).

38. George Musser, "Schrodinger's Zombie: Adam Brown at the 6th FQXi Meeting," *FQXi Blog*, September 6, https://fqxi.org/community /forum/topic/3345.

39. Larissa Albantakis, "Unfolding the Substitution Argument," *Conscious(ness) Realist*, September 14, 2020, https://www.conscious nessrealist.com/unfolding-argument-commentary.

40. Scott Aaronson, "Why I Am Not an Integrated Information Theorist (Or, the Unconscious Expander)," *Shtetl Optimized: The Blog of Scott Aaronson*, https://scottaaronson.blog/?p=1799.

41. Thomas Metzinger, *Being No One: The Self-Model Theory of Subjectivity* (Cambridge, MA: MIT Press, 2004).

42. Hakwan Lau and David Rosenthal, "Empirical Support for Higher-Order Theories of Conscious Awareness," *Trends in Cognitive Sciences* 15, no. 8 (2011): 365–373.

43. Michael S. A. Graziano and Taylor W. Webb, "The Attention Schema Theory: A Mechanistic Account of Subjective Awareness," *Frontiers in Psychology* (2015): 500.

44. Aaronson, "Why I Am Not an Integrated Information Theorist."

45. Matteo Grasso, Andrew M. Haun, and Giulio Tononi. "Of Maps and Grids," *Neuroscience of Consciousness*, no. 2 (2021): niab022.

46. Baars, "Global Workspace Theory of Consciousness."

47. Graziano and Webb, "The Attention Schema Theory."

48. Dennett, *Consciousness Explained*.

49. Nitasha Tiku, "The Google Engineer Who Thinks the Company's AI Has Come to Life," *New York Times*, June 11, 2022, https://www

.washingtonpost.com/technology/2022/06/11/google-ai-lamda
-blake-lemoine/.

50. Robert Miles, Twitter post, June 12, 2022, https://twitter.com/rob
ertskmiles/status/1536039724162469889.

51. Rohit Krishnan, Twitter post, June 12, 2022, https://twitter.com
/krishnanrohit/status/1536050803055656961.

52. Emily M. Bender, Timnit Gebru, Angelina McMillan-Major, and
Shmargaret Shmitchell, "On the Dangers of Stochastic Parrots: Can
Language Models Be Too Big?," *Proceedings of the 2021 ACM Conference on Fairness, Accountability, and Transparency* (2021): 610–
623.

53. Tiku, "The Google Engineer Who Thinks the Company's AI Has
Come to Life."

54. Blake Lemoine, Twitter post, June 13, 2022, https://twitter.com/ca
jundiscordian/status/1536503474308907010?lang=en.

55. Gilbert Ryle, *The Concept of Mind* (Chicago: University of Chicago
Press, 1949).

CHAPTER 7: THE TALE OF ZOMBIE DESCARTES

1. David Chalmers, *The Conscious Mind: In Search of a Theory of Conscious Experience* (Oxford: Oxford University Press, 1996).

2. Richard Dawkins, *The Selfish Gene* (Oxford: Oxford University
Press, 1976).

3. Joseph Levine, "Materialism and Qualia: The Explanatory Gap,"
Pacific Philosophical Quarterly 64, no. 4 (1983): 354–361.

4. Frank Jackson, "What Mary Didn't Know," *Journal of Philosophy* 83,
no. 5 (1986): 291–295.

5. Saul A. Kripke, "Naming and Necessity," in *Semantics of Natural
Language* (Dordrecht, Netherlands: Springer, 1972), 253–355.

6. Thomas Nagel, "What Is It Like to Be a Bat?" *Philosophical Review*
83, no. 4 (1974): 435–450.

7. Daniel C. Dennett, *Intuition Pumps and Other Tools for Thinking*
(New York: W. W. Norton & Company, 2013).

8. David Chalmers, "Does Conceivability Entail Possibility?" *Conceivability and Possibility* 145 (2002): 200.

9. Ned Block and Robert Stalnaker, "Conceptual Analysis, Dualism, and the Explanatory Gap," *Philosophical Review* 108, no. 1 (1999): 1–46.

10. Eric Marcus, "Why Zombies Are Inconceivable," *Australasian Journal of Philosophy* 82, no. 3 (2004): 477–490.

11. David Chalmers, "The Two-Dimensional Argument Against Materialism," in *The Oxford Handbook of the Philosophy of Mind*, ed. B. McLaughlin (Oxford: Oxford University Press, 2006).

12. Todd C. Moody, "Conversations with Zombies," *Journal of Consciousness Studies* 1, no. 2 (1994): 196–200.

13. Robert Kirk, *Zombies and Consciousness* (Oxford: Clarendon Press, 2005).

14. Nigel J. T. Thomas, "Zombie Killer. Toward a Science of Consciousness," in *Toward a Science of Consciousness II: The Second Tucson Discussions and Debates*, ed. S. R. Hameroff, A. W. Kaszniak, and A. C. Scott (Cambridge, MA: MIT Press, 1998), 171–177.

15. David Chalmers, "The Content and Epistemology of Phenomenal Belief," *Consciousness: New Philosophical Perspectives* 220 (2003): 271.

16. Katalin Balog, "Conceivability Arguments or the Revenge of the Zombies," *The Paideia Archive: Twentieth World Congress of Philosophy* 35 (1998): 34–45.

17. David Chalmers, "The Content and Epistemology of Phenomenal Belief."

18. René Descartes, "Meditations on First Philosophy," in *The Philosophical Works of Descartes*, trans. Elizabeth S. Haldane (Cambridge, UK: Cambridge University Press, 1978).

19. René Descartes, "Discourse on the Method," Project Gutenberg eBook, trans. John Veitch, 1995, https://www.gutenberg.org/files/59/59-h/59-h.htm.

20. Julietta Rose, "Is Chalmers's Zombie Argument Self-Refuting? And How," *Binghamton Journal of Philosophy* 1, no. 1 (2013): 105–132.

Notes

CHAPTER 8: THE PRINCESS AND THE PHILOSOPHER

1. Jonathan Bennett, "Correspondence Between René Descartes and Princess Elisabeth of Bohemia," 2009, https://www.earlymodern texts.com/assets/pdfs/descartes1643_1.pdf.
2. Ibid., 1.
3. Ibid., 2–4.
4. Ibid., 4.
5. Ibid., 7.
6. Ibid., 8.
7. Erik-Jan Bos, "Princess Elizabeth of Bohemia and Descartes' Letters (1650–1665)," *Historia Mathematica* 37, no. 3 (2010): 485–502.
8. Ibid., 9.

CHAPTER 9: CONSCIOUSNESS AND SCIENTIFIC INCOMPLETENESS

1. Alan Turing, "Computing Machinery and Intelligence" *Mind* 49 (1950): 433–460, https://redirect.cs.umbc.edu/courses/471/papers /turing.pdf, 6.
2. Stephen Wolfram, "A Class of Models with the Potential to Represent Fundamental Physics," arXiv preprint 2004.08210 (2020).
3. Jolly Mathen, "On the Inherent Incompleteness of Scientific Theories," *Activitas Nervosa Superior* 53, no. 1 (2011): 44–100.
4. Thomas Nagel, *The View from Nowhere* (Oxford: Oxford University Press, 1989).
5. Thomas Breuer, "The Impossibility of Accurate State Self-Measurements," *Philosophy of Science* 62, no. 2 (1995): 197–214.
6. Karl Svozil, *Physical (A) Causality: Determinism, Randomness and Uncaused Events* (Cham: Springer Nature, 2018).
7. Lawrence M. Krauss, *A Universe from Nothing: Why There Is Something Rather Than Nothing* (New York: Simon & Schuster, 2012).
8. Jim Holt, *Why Does the World Exist? An Existential Detective Story* (New York: W. W. Norton & Company, 2012).

9. John Horgan, *The End of Science: Facing the Limits of Knowledge in the Twilight of the Scientific Age* (New York: Basic Books, 2015).

10. Stephen W. Hawking, *A Brief History of Time: From the Big Bang to Black Holes* (New York: Bantam Books, 1988).

11. Stephen Hawking, "Gödel and the End of Physics," March 8, 2002, http://yclept.ucdavis.edu/course/215c.S17/TEX/GodelAndEnd OfPhysics.pdf.

12. John C. Sommerer and Edward Ott, "Intermingled Basins of Attraction: Uncomputability in a Simple Physical System," *Physics Letters* A214, no. 5–6 (1996): 243–251.

13. Jens Eisert, Markus P. Müller, and Christian Gogolin, "Quantum Measurement Occurrence Is Undecidable," *Physical Review Letters* 108, no. 26 (2012): 260501.

14. Alex Churchill, Stella Biderman, and Austin Herrick, "Magic: The Gathering Is Turing Complete," arXiv preprint1904.09828 (2019).

15. Toby Cubitt, David Perez-Garcia, and Michael M. Wolf, "Undecidability of the Spectral Gap," *Nature* 528 (2015): 207–211.

16. Toby S. Cubitt, "A Note on the Second Spectral Gap Incompleteness Theorem," arXiv preprint 2105.09854 (2021), 8.

17. Douglas Hofstadter, *Gödel, Escher, Bach* (New York: Basic Books, 1979).

18. Douglas Hofstadter, *I Am a Strange Loop* (New York: Basic Books, 2007).

19. Roger Penrose, *The Emperor's New Mind* (Oxford: Oxford University Press, 1989).

20. David J. Chalmers, "Minds, Machines, and Mathematics," *Psyche* 2, no. 9 (1995): 117–118.

21. John R. Lucas, "Minds, Machines and Gödel," *Philosophy* 36, no. 137 (1961): 112–127.

22. Roger Penrose, *Shadows of the Mind: A Search for the Missing Science of Consciousness* (Oxford: Oxford University Press, 1994).

23. Paul Benacerraf, "God, the Devil, and Gödel," *Monist* 51, no. 1 (1967): 9–32.

24. Panu Raattkainen, "On the Philosophical Relevance of Godel's

Incompleteness Theorems," *Revue Internationale de Philosophie* 4 (2005): 513–534.

25. Friedrich A. Hayek, *The Sensory Order: An Inquiry into the Foundations of Theoretical Psychology* (Chicago: University of Chicago Press, 1999), 194.

26. Ludwig Van den Hauwe, "Hayek, Gödel, and the Case for Methodological Dualism," *Journal of Economic Methodology* 18, no. 4 (2011): 387–407, 395.

27. Colin McGinn, *The Mysterious Flame: Conscious Minds in a Material World* (New York: Basic Books, 1999).

28. Colin McGinn, "Can We Solve the Mind-Body Problem?," *Mind* 98, no. 391 (1989): 349–366, 353.

CHAPTER 10: HOW SCIENCE GOT ITS SCALE

1. Richard Klavans and Kevin W. Boyack, "Toward a Consensus Map of Science," *Journal of the American Society for Information Science and Technology* 60, no. 3 (2009): 455–476.

2. David H. Hubel and Torsten N. Wiesel, "Receptive Fields, Binocular Interaction and Functional Architecture in the Cat's Visual Cortex," *Journal of Physiology* 160, no. 1 (1962): 106.

3. Vernon B. Mountcastle, "The Columnar Organization of the Neocortex," *Brain: A Journal of Neurology* 120, no. 4 (1997): 701–722.

4. Daniel P. Buxhoeveden and Manuel F. Casanova, "The Minicolumn Hypothesis in Neuroscience," *Brain* 125, no. 5 (2002): 935–951.

5. Apostolos P. Georgopoulos, Andrew B. Schwartz, and Ronald E. Kettner, "Neuronal Population Coding of Movement Direction," *Science* 233, no. 4771 (1986): 1416–1419.

6. Jerry A. Fodor, "Special Sciences (Or: The Disunity of Science as a Working Hypothesis)," *Synthese* 28, no. 2 (1974): 97–115.

7. Erik Hoel, "Agent Above, Atom Below: How Agents Causally Emerge from Their Underlying Microphysics," in *Wandering Towards a Goal*, ed. Anthony Aguirre, Brendan Foster, Zeeya Merali (Cham: Springer, 2018), 63–76.

8. Richard Gallagher and Tim Appenzeller, "Beyond Reductionism," *Science* 284, no. 5411 (1999): 79–79.

9. Jaegwon Kim, *Mind in a Physical World: An Essay on the Mind-Body Problem and Mental Causation* (Cambridge, MA: MIT Press, 1998).

10. Ned Block, "Do Causal Powers Drain Away?," *Philosophy and Phenomenological Research* 67, no. 1 (2003): 133–150.

11. Thomas D. Bontly, "The Supervenience Argument Generalizes," *Philosophical Studies* 109, no. 1 (2002): 75–96.

12. Ismail Zaitoun, Karen M. Downs, Guilherme J. M. Rosa, and Hasan Khatib, "Upregulation of Imprinted Genes in Mice: An Insight into the Intensity of Gene Expression and the Evolution of Genomic Imprinting," *Epigenetics* 5, no. 2 (2010): 149–158.

13. Karl Deisseroth, "Optogenetics," *Nature Methods* 8, no. 1 (2011): 26–29.

14. Simone Sarasso et al., "Consciousness and Complexity During Unresponsiveness Induced by Propofol, Xenon, and Ketamine," *Current Biology* 25, no. 23 (2015): 3099–3105.

15. Carlo Rago, Bert Vogelstein, and Fred Bunz, "Genetic Knockouts And Knockins In Human Somatic Cells," *Nature Protocols* 2, no. 11 (2007): 2734–2746.

16. Jason Grossman and Fiona J. Mackenzie, "The Randomized Controlled Trial: Gold Standard, or Merely Standard?," *Perspectives in Biology and Medicine* 48, no. 4 (2005): 516–534.

17. Nabil Guelzim, Samuele Bottani, Paul Bourgine, and François Képès, "Topological and Causal Structure of the Yeast Transcriptional Regulatory Network." *Nature Genetics* 31, no. 1 (2002): 60–63.

18. Marinka Zitnik, Rok Sosič, Marcus W. Feldman, and Jure Leskovec, "Evolution of Resilience in Protein Interactomes Across the Tree of Life," *Proceedings of the National Academy of Sciences* 116, no. 10 (2019): 4426–4433.

19. Judea Pearl, *Causality: Models, Reasoning, and Inference* (Cambridge, UK: Cambridge University Press, 2009).

20. Ibid., 415.

21. Judea Pearl and Dana Mackenzie, *The Book of Why: The New Science of Cause and Effect* (New York: Basic Books, 2018).

22. Nancy Cartwright, *The Dappled World: A Study of the Boundaries of Science* (Cambridge, UK: Cambridge University Press, 1999).

23. Murray Gell-Mann, "What Is Complexity?" *Complexity* 1 (1995): 16–19.

24. David Hume, *An Enquiry Concerning Human Understanding*, ed. Jonathan Bennett (2017), 38, https://www.earlymoderntexts.com/assets/pdfs/hume1748.pdf.

25. Ibid.

26. David Lewis, "Causation," *Journal of Philosophy* 70, no. 17 (1974): 556–567.

27. David Lewis, *Philosophical Papers Volume I* (Oxford: Oxford University Press, 1983).

28. Stephen Yablo, "Mental Causation," *Philosophical Review* 101, no. 2 (1992): 245–280.

29. Christian List and Peter Menzies, "Nonreductive Physicalism and the Limits of the Exclusion Principle," *Journal of Philosophy* 106, no. 9 (2009): 475–502.

30. Christian List, "Free Will, Determinism, and the Possibility of Doing Otherwise," *Noûs* 48, no. 1 (2014): 156–178.

31. Christopher Hitchcock, "Probabilistic Causation," in *The Stanford Encyclopedia of Philosophy* (Spring 2021 Edition), https://plato.stanford.edu/archives/spr2021/entries/causation–probabilistic.

32. Larissa Albantakis, William Marshall, Erik Hoel, and Giulio Tononi, "What Caused What? A Quantitative Account of Actual Causation Using Dynamical Causal Networks," *Entropy* 21, no. 5 (2019): 459.

33. Fernando E. Rosas et al., "Reconciling Emergences: An Information-Theoretic Approach to Identify Causal Emergence in Multivariate Data," *PLoS Computational Biology* 16, no. 12 (2020): e1008289.

34. Thomas F. Varley and Erik Hoel, "Emergence as the Conversion of Information: A Unifying Theory," *Philosophical Transactions of the Royal Society A* 380, no. 2227 (2022): 20210150.

35. Pedro A. M. Mediano et al., "Greater Than the Parts: A Review of

the Information Decomposition Approach to Causal Emergence," *Philosophical Transactions of the Royal Society A*380, no. 2227 (2022): 20210246.

36. Erik P. Hoel, Larissa Albantakis, and Giulio Tononi, "Quantifying Causal Emergence Shows That Macro Can Beat Micro," *Proceedings of the National Academy of Sciences* 110, no. 49 (2013): 19790–19795.

37. Erik P. Hoel, Larissa Albantakis, William Marshall, and Giulio Tononi, "Can the Macro Beat the Micro? Integrated Information Across Spatiotemporal Scales," *Neuroscience of Consciousness*, no. 1 (2016).

38. Scott Aaronson, "Higher Level Causation Exists (But I Wish It Didn't)," *Shtetl Optimized: The Blog of Scott Aaronson*, https://scott aaronson.blog/?p=3294.

39. Frederick Eberhardt and Lin Lin Lee, "Causal Emergence: When Distortions in a Map Obscure the Territory," *Philosophies* 7 , no. 2 (2022): 30.

40. Renzo Comolatti and Erik Hoel, "Causal Emergence Is Widespread Across Measures of Causation," arXiv preprint 2202.01854 (2022).

41. P. W. Anderson, "More Is Different: Broken Symmetry and the Nature of the Hierarchical Structure of Science," *Science* 177, no. 4047 (1972): 393–396.

42. Steven Strogatz, et al., "Fifty Years of 'More Is Different,'" *Nature Reviews Physics* 4, no. 8 (2022): 508–510.

43. Richard W. Hamming, "Error Detecting and Error Correcting Codes," *Bell System Technical Journal* 29, no. 2 (1950): 147–160.

44. Erik Hoel, "When the Map Is Better Than the Territory," *Entropy* 19, no. 5 (2017): 188.

45. Giulio Tononi, Olaf Sporns, and Gerald M. Edelman, "Measures of Degeneracy and Redundancy in Biological Networks," *Proceedings of the National Academy of Sciences* 96, no. 6 (1999): 3257–3262.

46. Comolatti and Hoel, "Causal Emergence Is Widespread Across Measures of Causation."

47. Paul C. W. Davies, "Emergent Biological Principles and the Com-

putational Properties of the Universe," arXiv preprint astro-ph /0408014 (2004).

48. David J. Chalmers, "Strong and Weak Emergence," in *The Re-emergence of Emergence* (2006), 244–256.

49. Comolatti and Hoel, "Causal Emergence Is Widespread Across Measures of Causation."

50. A. Einstein, "Investigations on the Theory of the Brownian Movement," *Annalen der Physik* 17 (1905): 549.

51. David Colquhoun and A. G. Hawkes, "On the Stochastic Properties of Single Ion Channels," *Proceedings of the Royal Society of London, Series B, Biological Sciences* 211, no. 1183 (1981): 205–235.

52. Stephen Wolfram, "A Class of Models with the Potential to Represent Fundamental Physics," arXiv preprint 2004.08210 (2020).

53. Lee Smolin, *The Trouble with Physics: The Rise of String Theory, the Fall of a Science, and What Comes Next* (Boston: Houghton Mifflin Company, 2007).

54. S. Hossenfelder and T. Palmer, "Rethinking Superdeterminism," *Frontiers in Physics* 8 (2020): 139.

55. David H. Wolpert, "Physical Limits of Inference," *Physica D: Nonlinear Phenomena* 237, no. 9 (2008): 1257–1281.

56. John Conway and Simon Kochen, "The Strong Free Will Theorem," *Notices of the AMS* 56, no. 2 (2009): 226–232.

57. Pearl, *Causality: Models, Reasoning, and Inference*, 420.

58. Ross Griebenow, Brennan Klein, and Erik Hoel, "Finding the Right Scale of a Network: Efficient Identification of Causal Emergence Through Spectral Clustering," arXiv preprint 1908.07565 (2019).

59. Brennan Klein and Erik Hoel, "The Emergence of Informative Higher Scales in Complex Networks," *Complexity* 2020 (2020).

60. Brennan Klein et al., "Evolution and Emergence: Higher Order Information Structure in Protein Interactomes Across the Tree of Life," *Integrative Biology* 13, no. 12 (2021): 283–294.

CHAPTER 11: THE SCIENTIFIC CASE FOR FREE WILL

1. A. Will Crescioni et al., "Subjective Correlates and Consequences of Belief in Free Will," *Philosophical Psychology* 29, no. 1 (2016): 41–63.
2. Daniel C. Dennett, *Elbow Room, New Edition: The Varieties of Free Will Worth Wanting* (Cambridge, MA: MIT Press, 2015).
3. Jerry Fodor, "Making Mind Matter More," *Philosophical Topics* 17: 59–80, 77.
4. Erik P. Hoel, Larissa Albantakis, and Giulio Tononi, "Quantifying Causal Emergence Shows That Macro Can Beat Micro," *Proceedings of the National Academy of Sciences* 110, no. 49 (2013): 19790–19795.
5. Hans H. Kornhuber and Lüder Deecke, "Hirnpotentialänderungen bei Willkürbewegungen und passiven Bewegungen des Menschen: Bereitschaftspotential und reafferente Potentiale," *Pflüger's Archiv für die Gesamte Physiologie des Menschen und der Tiere* 284, no. 1 (1965): 1–17.
6. Benjamin Libet, "Unconscious Cerebral Initiative and the Role of Conscious Will in Voluntary Action," *Behavioral and Brain Sciences* 8, no. 4 (1985): 529–539.
7. Benjamin Libet, "Do We Have Free Will?" *Journal of Consciousness Studies* 6, no. 8–9 (1999): 47–57.
8. Aaron Schurger, Jacobo D. Sitt, and Stanislas Dehaene, "An Accumulator Model for Spontaneous Neural Activity Prior to Self-Initiated Movement," *Proceedings of the National Academy of Sciences* 109, no. 42 (2012): E2904–E2913.
9. Galen Strawson, "The Impossibility of Moral Responsibility," *Philosophical Studies: An International Journal for Philosophy in the Analytic Tradition* 75, no. 1/2 (1994): 5–24.
10. Larissa Albantakis, Francesco Massari, Maggie Beheler-Amass, and Giulio Tononi. "A Macro Agent and Its Actions," in *Top-Down Causation and Emergence* (Cham: Springer, 2021), 135–155.

Notes

11. Peter Van Inwagen, *An Essay on Free Will* (Oxford: Clarendon Press, 1983).

12. *The Consolation of Philosophy of Boethius*, trans. H. R. James, https://www.gutenberg.org/files/14328/14328.txt.

13. Stephen Wolfram, *A New Kind of Science* (Champaign, IL: Wolfram Media, 2002), 739–741, https://www.wolframscience.com/nks/.

14. Richard Taylor, "Fatalism," *Philosophical Review* 71 (1962): 56–66.

15. John T. Saunders, "Fatalism and the Logic of 'Ability,'" *Analysis* 24, no. 1 (1963): 24–24.

16. David F. Wallace, *Fate, Time, and Language* (New York: Columbia University Press, 2011).

About the Author

Erik Hoel is one of the *Forbes* 30 Under 30 in science and also a New York City Emerging Writers Fellow. His writing has appeared in *The Atlantic*, and his nonfiction was selected as notable by the Best American Essays series. Hoel received his PhD in neuroscience at the University of Wisconsin–Madison and has been a postdoctoral researcher at Columbia University, a visiting scholar at the Institute for Advanced Study at Princeton, and a research professor at Tufts University. He grew up in his mother's independent bookstore, The Jabberwocky, becoming a lifelong writer and book lover. His first novel, *The Revelations*, was published in 2021, and he currently runs a popular Substack, *The Intrinsic Perspective*. He lives on Cape Cod in Massachusetts.